P9-CQW-382

Maintaining and Troubleshooting HPLC Systems

Maintaining and Troubleshooting HPLC Systems

A USER'S GUIDE

DENNIS J. RUNSER, Ph.D., C.P.C.

A WILEY-INTERSCIENCE PUBLICATION

JOHN WILEY & SONS

New York • Chichester • Brisbane • Toronto • Singapore

Library of Congress Cataloging in Publication Data:

Runser, Dennis J

Maintaining and troubleshooting HPLC systems.

"A Wiley-Interscience publication."
Includes index.
1. Liquid chromatography—Handbooks, manuals, etc.
I. Title.

QD79.C454R86 543'.0894 80-25444
ISBN 0-471-06479-3

Printed in the United States of America

10 9 8 7 6 5 4 3 2

To

Robert and David

Preface

High performance liquid chromatography (HPLC) has become one of the most important and fastest growing techniques in the analytical laboratory. Although it is only about 10 years old, it now rivals gas chromatography in its speed and efficiency of separation. Numerous papers and texts have been published on HPLC, covering its theory, instrumentation, practice, and application. However, until now no text has been published covering exclusively an equally important aspect of this methodology, namely, the maintenance and troubleshooting of HPLC systems.

The purpose of this book is to provide the HPLC user with the first comprehensive manual for troubleshooting and maintaining HPLC systems. It is not intended to describe chromatographic theory and its application to maximizing separation nor to describe each HPLC component in relationship to application. Furthermore, the text is not simply a collection of instrument manufacturers' user manuals. Its objective is to describe how the user can maintain the HPLC system in operating condition, what to look for and do to prevent and solve problems, and what should be done before calling a service representative.

The text is prepared and arranged as a guide and working manual the chromatographer can use in the laboratory. It is organized into chapters which basically represent the typical components of a liquid chromatographic system. Each of these chapters describes a basic element of the HPLC system in terms of advantages and disadvantages to its care and operation. Recommendations are made for establishing good operating procedures which will minimize downtime and ultimately maximize method development and analysis time. These sections are written so that the chromatographer may use the book as a handy guide right in the laboratory.

A chapter is also devoted to laboratory safety pertaining to the practice of HPLC. The impact of OSHA and EPA has made it mandatory for the chromatographer to be aware of safe laboratory protocol when using HPLC equipment, solvents, samples, and effluents. This chapter addresses these concerns and presents preventative measures and procedures.

Finally, the appendices include useful guides in several key maintenance areas. A comprehensive troubleshooting guide describing symptoms, potential causes, and recommended solutions to system problems is included. A procedure for calibrating a strip chart recorder, tools necessary to perform routine maintenance and troubleshooting, and aspects to consider before calling the local service representative are presented.

The successful chromatographer is not only one who can develop HPLC separations and analytical procedures, but one who can also maintain and troubleshoot his $3,000–$30,000 HPLC unit. With the cost of service calls running $25–$50 per hour plus travel and parts, the cost of this text can well pay for itself for each service call prevented.

DENNIS J. RUNSER

Stilwell, Kansas
January 1981

Contents

Maintaining and Troubleshooting HPLC Systems

1
Introduction

1.1 BACKGROUND

High performance liquid chromatography (HPLC) has become the fastest growing analytical technique in the last 10 years. This is evident by the increased number of technical sessions and papers on HPLC at national and local conferences, symposia, and seminars, by the expanded coverage it has been receiving in technical journals as well as in journals devoted exclusively to HPLC, by the rapid growth of manufacturers of HPLC equipment and supplies, and by the projected purchases of HPLC equipment (1–2).

The basic theory behind high performance liquid chromatography is not new, but it was not until around 1969 that HPLC as we now know it was developed. This development introduced the effective use of small diameter packing materials and columns which allowed the chromatographer to perform separations faster and with greater resolution than had previously been attainable.

These advances put liquid chromatography into a practicing methodology similar to gas chromatography (GC), which had become the most popular chromatographic method of the sixties. Since HPLC is not limited to sample volatility and thermal stability as is GC, nearly all classes of organic compounds can be separated by this highly efficient and faster analytical technique, which utilizes the basic liquid chromatography principles demonstrated by Martin and Synge in 1941 (3). Thus the practice of liquid chromatography went from open column, gravity fed systems utilizing effluent collection, followed by various analytical techniques, such as evaporation to dryness using histogram plots and subsequent dilution and/or reactions followed by ultraviolet (UV)-visible and infrared spectroscopy, to a totally instrumental methodology.

The decreases in column packing diameters produced proportional increases in solvent flow resistance. As a result, constant flow and constant pressure liquid pumping systems were developed and introduced. Sample introduction by simply adding the sample to the head of the column was no longer practical, and liquid syringe and valve injectors were added for sample introduction. The efficiency of HPLC allowed very small samples to be used, and the classical methods of collecting effluents for further analysis were replaced by using low dead volume detectors coupled with strip chart recorders. Yet the problems associated with the handling, care, and troubleshooting of an intricate analytical instrument were not avoided.

1.2 SCOPE OF THE TEXT

Numerous texts have been written on HPLC (4–9). They describe its theory, applications, and instrumentation. These aspects of HPLC will not be discussed or reviewed here.

However, up to now there has not been a single text entirely devoted to the maintenance and troubleshooting of HPLC systems. This subject has so far only been treated as an individual chapter in HPLC texts (5, 8) or as part of a text (10). The scope of this book is to discuss only this single aspect of HPLC.

The subject is presented from a systems approach, that is, the maintenance and troubleshooting of the HPLC system are treated through each of its individual components. The HPLC model to be used is illustrated in Figure 1.1.

The text is also intended to be a user's manual to solve maintenance and troubleshooting problems. Therefore extensive use is made of cross references, and an in-depth index is provided. The text is not intended to be a catalog or showcase for every instrument currently available, but rather a guide using examples of different manufacturers' instruments and models to illustrate various maintenance and troubleshooting techniques and to present examples that may be extrapolated to other models.

There is a familiar saying among users of instruments and equipment in all

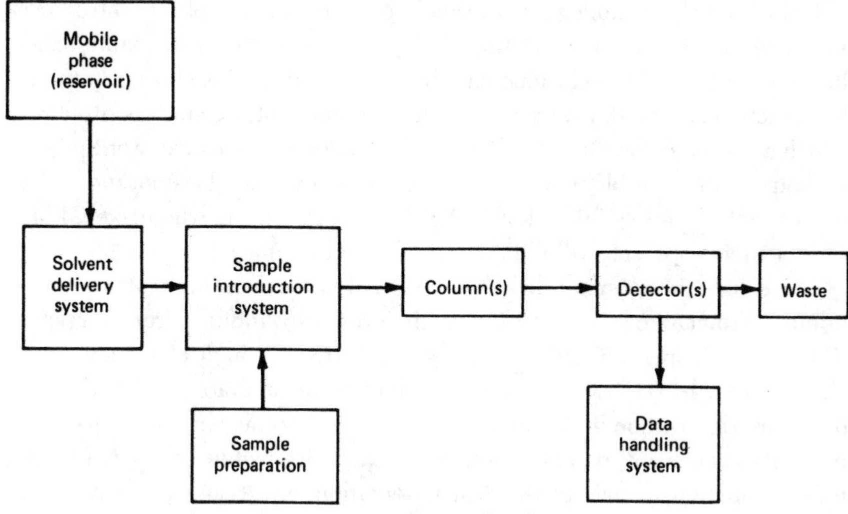

Figure 1.1 HPLC system diagram.

fields, "If all else fails, read the directions." This book is not an attempt to rewrite manufacturers' user manuals. These already exist. Unfortunately so many analysts, especially experienced chromatographers, do not read the user manual until a problem has already surfaced. This can be a serious mistake. The user manual is an important reference toward the problem-free and safe use of a particular manufacturer's piece of equipment. Not all instruments or systems are alike. This text is intended to supplement these manuals and to inform the user on what to do and look for before calling the service representative. Together, the instrument manuals and this text should provide the chromatographer with enough information to operate relatively trouble-free.

1.3 THE PREVENTATIVE MAINTENANCE ATTITUDE

Although the chromatographer can use this book and other aids in maintaining the HPLC system, one underlying aspect is most critical to the user's success. It cannot be taught but must be learned through conscientious experience. I call this the chromatographers' preventative maintenance attitude (CPMA).

Just as the chromatographer routinely prepares mobile phase, introduces samples, and so on, so must the concept of prevention and maintenance become routine. If the chromatographer cannot adopt this concept into his/her practicing everyday attitude, he/she will encounter countless problems which will create downtime, frustration, and poor performance. For the user without CPMA, problems will appear as surprises. This does not mean that one cannot be successful without CPMA, but those users who make CPMA part of their approach will obtain superior performance.

There have been many definitions of HPLC, such as high performance liquid chromatography (HPLC), high pressure liquid chromatography (HPLC), high speed liquid chromatography (HSLC), high efficiency liquid chromatography (HELC), and simply liquid chromatography (LC). There is one more description which in terms of troubleshooting and maintenance is particularly relevant, namely, high price liquid chromatography (HPLC). In today's inflationary market this instrumentation and its use are expensive. More costly still is the user's time. If for no other reason, CPMA can reduce user costs while maintaining the HPLC system and its components. The more ingrained this attitude is in the user, the more efficient and valuable he/she will be to his/her employer, business, and research.

I have taught scores of analysts in the art of practicing HPLC, and the one most singly defeating aspect I have observed is the analyst's frustrations over the system and equipment problems he/she encounters. I cannot stress enough the value of having a maintenance and troubleshooting attitude as everyday practice for the successful use of the material presented in the chapters that follow. It is as simple as reading the manual first.

REFERENCES

1 E. J. Thomas, Jr., *Research/Development*, 24–26 (February 1978).

2 J. M. Petruzzi, *Analytical Chemistry*, **51**(1) 86A (1979).

3 A. J. P. Martin and R. L. M. Synge, *J. Biochemistry*, **35**, 1358 (1941).

4 *Modern Practice of Liquid Chromatography*, J. J. Kirkland, Ed., Wiley-Interscience, John Wiley & Sons, Inc., New York, 1971.

5 E. L. Johnson and R. Stevenson, *Basic Liquid Chromatography*, Varian Associates, Palo Alto, CA, 1978.

6 S. G. Perry, R. Amos and P. I. Brewer, *Practical Liquid Chromatography*, Plenum Press, New York, 1972.

7 P. R. Brown, *High Pressure Liquid Chromatography: Biochemical and Biomedical Applications*, Academic Press, Inc., New York, 1973.

8 L. R. Synder and J. J. Kirkland, *Introduction to Modern Liquid Chromatography*, 2nd. Edition, Wiley-Interscience, John Wiley & Sons, Inc., New York, 1979.

9 K. Wessely and K. Zech, *High Performance Liquid Chromatography in Pharmaceutical Analyses,* Hewlett-Packard GmbH, Böblingen, Germany, 1979.

10 J. Q. Walker, M. T. Jackson, Jr., and J. B. Maynard, *Chromatographic Systems: Maintenance and Troubleshooting*, Academic Press, New York, 1977.

2
The Mobile Phase

2.1 INTRODUCTION

In considering liquid chromatography from the systems approach, the first component to examine is the mobile phase and solvent reservoir. The development of HPLC over the last 10 years has clearly placed increased emphasis on the qualities and handling of the mobile phase. As in all forms of chromatography, users know that the quality and proper handling of the carrier (gas, mobile phase, developing solvent) are as critical as any other part of the chromatographic system. Yet it is all too common for the user to take the carrier for granted and stress the more technically significant system components, such as the solvent delivery system, the detector, etc. It is very simple to take a bottle of labled reagent grade solvent, pour it into the solvent reservoir, and start the chromatograph. However, when a problem develops with the mobile phase, it may not be easily or immediately identified, and once it is, the entire system is already affected, causing considerable change over time.

The most frequently encountered problems can be avoided or reduced by careful handling of the choice of solvent(s), the mobile phase preparation, the solvent reservoir, and the changing of the mobile phase.

2.2 CHOICE OF SOLVENT(S)

The choice of solvent is most frequently made by selecting one of the four major areas of HPLC (liquid-solid, liquid-liquid, ion-exchange, and steric exclusion) that is expected to provide the desired separation. The strategy for selecting the right solvent system to achieve the desired separation has been covered by other authors (1–7). However, the choice based on preventing system problems is described in the following paragraphs.

2.2.1 System Compatibility

The HPLC system itself needs to be considered when separation solvents are selected. The solvent system must be miscible with the previously used mobile phase. If it is not, an intermediate solvent needs to be considered first, one that is soluble with both the previous mobile phase and the new one. This will require flushing the system and introducing the new solvent, which may take up to 30 min. If the solvents are miscible, some system rinsing time will still be needed, but significantly less, such as 5–15 min.

Godfrey (8) has developed a method for predicting the miscibility of various organic solvent pairs. It is based on studying the partitioning of a series of organic solvents from low to high lipophilicity and assigning a numerical value to each solvent, called the miscibility number (M), where M has values from 1 to 31 and each value represents a uniform interval between miscibility classes which have quantitative significance as a measure of lipophilicity. This is a relative procedure for predicting miscibility. It uses the following criteria between solvent pairs:

1 When $M \leq 15$, the solvent pair is miscible in all proportions at 25°C.
2 When $M = 16$, the solvent pair has a critical solution temperature between 25°C and 75°C.
3 When $M \geq 17$, the solvent pair is generally immiscible.
4 When M has two values, the first value (usually less than 16) defines miscibility with solvents of high lipophilicity, and the second value (usually greater than 16) defines miscibility with solvents of low lipophilicity.

Table 2.1 presents miscibility numbers in relative decreasing order of polar-

Table 2.1 Miscibility Numbers for Some Typical HPLC Solvents

Solvent	Miscibility Number *(M)*
Acetic acid	14
Ethylene glycol	2
Glycerol	1
Formamide	3
Acetic anhydride	12, 19
Furfural	11, 17
Acetonitrile	11, 17
Methanol	12
Ethanol	14
Dimethyl formamide	12
1-Propanol	15
iso-Propanol	15
Methyl ethyl ketone	17
2-Butanol	16
1-Butanol	15
Cyclohexanol	16
Pyridine	16
Dimethyl sulfoxide	9
Methyl acetate	15, 17
Ethyl acetate	19
Dioxane	17
Tetrahydrofuran	17
Butyl acetate	22
Chloroform	19
Ethyl ether	23
1-Octanol	17
Trichloroethylene	20
Benzene	21
Chlorobenzene	21
Mestiylene	24
Toluene	23
Tetrachloroethylene	25
Isopropyl ether	26
p-Xylene	24

Table 2.1 (continued)

Solvent	Miscibility Number *(M)*
Carbon tetrachloride	24
Carbon disulfide	26
Cyclohexane	28
Decane	29
Octane	29
Hexane	29
iso-Octane	29
Heptane	29

(Reprinted with permission from CHEMTECH, 359–363 copyright 1972 by American Chemical Society.)

ity for typical organic solvents used in HPLC. Additional tables of miscibility of organic solvent pairs are published elsewhere (9).

Ignoring miscibility results in the forming of immiscible pools throughout the system, causing detector readout variations, drifts, and so on, which it may be very difficult to clean out. These pools can become entrapped in the system and pass through later at an inopportune time, ruining needed data.

When columns are changed, whether or not an intermediate flushing solvent is needed, simply removing the column and adding a piece of stainless steel tubing between the pump (or sample valve) and the detector will facilitate a quick system flush. Many pumps allow for quick draining from the solvent reservoir and have low dead volumes, so this may take only about 5 min.

When the same column is to be used, it can be disconnected and the pump and sample introduction part of the HPLC system can be purged quickly.

However, in either case the column will have to be purged with the new (or possibly intermediate) mobile phase. Usually 5–10 column volumes will be needed. It is important to note what mobile phase had last been used in the column (see Section 5.3.4). Once the entire system is running, the detector baseline can be used to indicate whether or not the system has stabilized by the absence of drifting and short term noise.

2.2.2 Column

When the mode of HPLC to be used is chosen, solvent compatibility with the column is important, particularly with certain chemically bonded liquid-liquid partition columns. For example, a Carbowax 400 on Corasil column (Waters Associates) which is bonded through an ester linkage will cleave with water or low molecular weight alcohols as the mobile phase. Most columns today use packings with surface-reacted or "permanently" bonded phases, and except for extremes in pH (see Section 5.3.1), they usually present no problem. For steric exclusion columns the type of solvent is critical with regard to column bed swelling, which will seriously affect the interstitial volume as well as the bed volume in the column itself.

Consideration of the column requires prior knowledge of the column packing, type of bonding, and so on. Ignoring this results in the creation of voids in the column, producing peak broadening and loss of resolution due to column deterioration.

2.2.3 Gradient Elution

With gradient elution systems at least two solvents (A and B) are used, and often a binary mixture which is varied 20–80% of its A/B ratio. Again miscibility and the ability to return to the desired starting condition are important. The larger the percent concentration variation, the longer the time needed to return to initial conditions. Ideally reversing the gradient and slowly returning to the starting conditions is preferred. This will maintain the column's lifetime the longest.

2.2.4 Solubility with the Sample(s)

Is the sample soluble in the mobile phase? If it is not, or if components of the sample are insoluble in the mobile phase, several problems may arise.

Precipitation may occur, clogging the syringe or sampling valve. Depending on the insoluble nature of the sample, it may damage the sampling valve by acting as grit and creating crossport scratches. These particulates can also build up at the head of the column, restricting the flow of the mobile phase and increasing the back pressure on the system. With HPLC systems, particularly those of limited pressure capability (3000 psi or less), this can create a problem of limited flow rate. Such a contaminant buildup can also cause spurious peaks (see Section 5.4.2).

Samples of limited solubility or materials dissolved in a very soluble solvent, such as MeOH, can precipitate out on the column. With subsequent injections the insoluble material can be redissolved and either partially carried through the column and then reprecipitated or eventually eluted, causing random and spurious peaks to be observed with the detector. This insolubility can also cause impurity buildup in detector cells as well as create drifting.

Although insolubility of the sample in the mobile phase may not create a problem, the chromatographer must be aware of its potential hazard. Preparation of the sample in the mobile phase itself is ideal in preventing this problem (see Section 4.2). In binary and tertiary systems direct solubility may be difficult to obtain. If this occurs, the material can be dissolved in the more soluble solvent and diluted to the mobile phase concentration with the other solvent(s). Another approach is to prepare a more concentrated solution in the more soluble solvent. Pipette a small aliquot into a volumetric flask, and bring it to volume with the mobile phase or evaporate the concentrate to dryness and then bring it to volume with the mobile phase.

The use of a different solvent for sample preparation will probably add to the void volume peak which can obscure early eluting peaks.

2.2.5 Solvent Quality

The quality of the solvent can be very critical. In the early days of HPLC (early 1970s) little emphasis was placed on the quality of the purchased solvent, other than whether it was reagent grade or spectroscopic grade. The obvious technical and practical grades were seldom used. However, chromatographers often experience differences in system repeatability with different manufacturers' solvents. This can be a problem to the extent of having to validate a specific procedure with another manufacturer's solvent(s) before changing sources. Differences in water content and organic impurities are the most common problems. A solvent simply labeled spectral or pesticide grade will not necessarily be free of contaminants for today's HPLC columns.

Knowing what stabilizers are present in the solvent can be critical. For example, chloroform usually contains up to 1.0% (v/v) methanol or ethanol, and tetrahydrofuran may contain butylated hydroxytoluene. In essence you already have a binary system in the "one" solvent itself. As a method is developed, note the grade, the concentration of the stabilizer(s) if present, and any other attributes of the solvents used for other chromatographers to attain reproducibility with your conditions. The use of a solvent with a stabilizer versus one without can reverse the elution pattern of two solutes. This can occur when using an HPLC system with stabilizer-free chloroform, and then changing to chloroform with a stabilizer. Ethyl acetate with H_2O usually already present can deactivate a liquid-solid column. This concentration of water can vary from bottle to bottle, changing k'. On the other hand, the lack of a stabilizer or the hygroscopic nature of a solvent can cause changes in the stock solvent, which can also change the k' of the system, as in solvents like dioxane and chloroform. Dangerous products can form, such as peroxides or phosgene respectively. Unstabilized chlorinated solvents can also degrade, creating free chloride ions, which can form hydrochloric acid which in turn will attack the HPLC system's stainless steel parts. These solvents are usually packaged under a nitrogen blanket.

2.2.6 Water

Water has become nearly the highest volume solvent used in HPLC. This is attributed to the ever-increasing use of reverse phase chromatography and, of course, ion-exchange and gel filtration systems. Since laboratory distilled water is considered acceptable for the average analytical laboratory, it is common to pay little attention to the impurities present in "distilled water." The need to pay attention to this aspect is dramatically illustrated in reverse phase systems where sensitive trace analyses are carried out and baseline noise occurs due to water containing organic contaminants.

Many analytical laboratories have plastic carboys of distilled water at the end of the laboratory bench. The water is usually generated with in-house laboratory stainless steel stills, small dedicated laboratory units, or plant systems. The carboys are filled regularly as needed.

One of the first problems with these containers is that no one cleans them and microbiological growth goes rampant. The multitude of such colonies of bacteria, mold and/or fungus in reverse phase, aqueous size exclusion (gel filtration), and ion-exchange systems will clog column interstitial spaces. The water standing overnight in the column may support such growth, increasing column back pressures. This can be prevented by cleaning the carboys, filtering the water through a 2 micron filter, or adding 0.02% sodium azide or acetonitrile (which in many reverse phase systems is already present) to the water.

The container material for storing the water also can become a problem.

With plastics, plasticizers will leach into the water and interfer with gradient reverse phase systems or contaminate the column. With glass, metal ions can leach from the surface, increasing water resistivity measurements when this is used as a criterion for water purity. However, glass leaching, which also occurs with glass distillation equipment, is usually not as much a problem as are organic impurities.

To avoid this, many chromatographers prepare or buy commercially available HPLC water, such as water from J. T. Baker or Burdick & Jackson. Simple commercial glass distillation equipment from manufacturers such as Corning and Barnsted can be very ample for the average HPLC laboratory. Using doubly distilled water by distilling the common laboratory supply is very adequate. The water can be checked periodically using the resistance measurement specification of 1 or 2 megohms as a measure of the still quality. These laboratory stills usually produce water at a rate of 1–4 liters/hr.

For gradient reverse phase there is no guarantee that volatile organics will not distill over. In this case several other cleanup procedures exist. The water can first be passed through a 2 ft C_{18} porous silica column as a cleanup step by running the water through the column overnight. The organics will be retained. Store the water in glass and not in plastic. Commercial water purification systems such as Millipore's Milli-Q/Milli-RO® systems are frequently used (Figure 2.1). In this case deionized water (preferred over potable water due to the resin lifetime) is fed first through a filter to remove particulates. The water passes then through a reverse osmosis cartridge to remove microorganisms, ionic contaminates, and pyrogens. Commonly, once through this system, the water passes to the Milli-Q® system (Figure 2.2) composed of a carbon cartridge to remove organic contaminants, a strong cation and a strong anion exchange column to remove metal ions, and then through a 0.22 micron particle trap. This system is advertised at producing 1.5 liters/min.

If impurities are suspected in the water, the following procedure can be used to check its purity (11):

1. Pump 100 ml of the water through a 2 mm ID × 61 cm C_{18} porous silica column.
2. Run a linear gradient from 0 to 100% methanol at 1 ml/min for 10 min and hold for 15 min with a UV detector in line.
3. If the UV baseline shift at 0.08 AUFS is less than 10% and very few

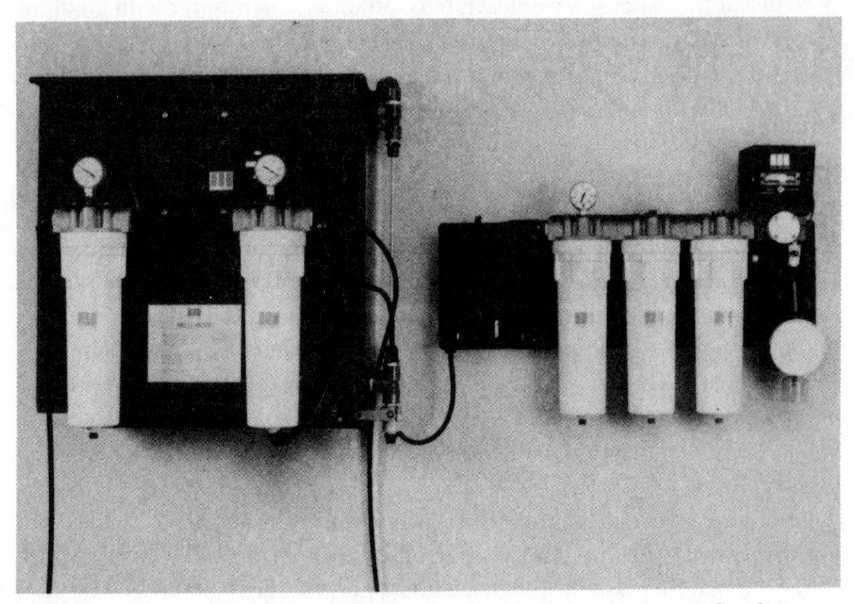

Figure 2.1 Millipore's Milli-Q/Milli-RO® water purification system (reprinted with permission of Millipore Corporation).

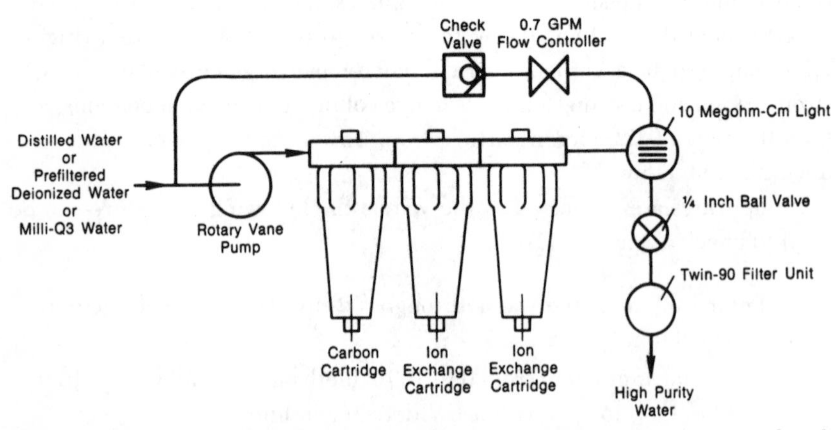

Figure 2.2 Schematic of Millipore's Milli-Q® water purification system (reprinted with permission of Millipore Corporation).

peaks of less than 3–5% full scale deflections are observed, the water is pure enough for most applications. With bad water ≥10 peaks can be observed to run off scale at 0.08 AUFS.

In considering the quality and choice of a solvent, Majors (10) gives a comprehensive table illustrating typical effects of impurities in HPLC solvents (see Table 2.2).

2.3 MOBILE PHASE PREPARATION

The preparation of the mobile phase is simple and relatively trouble-free. However, there are some problems which, if not avoided, can cause problems for the chromatographer.

2.3.1 Filtering and Degassing of Solvents

The first routine practice often neglected because the chromatographer is in a hurry is filtering the solvent before adding it to the solvent reservoir or solvent delivery system. Even if the HPLC system has filters routinely in line, this simple preventative maintenance practice can save future filter clogging, filter changes, and associated expense, and possibly downtime or system troubleshooting. However, whether or not filters are in the system and before the column, this is part of the CPMA which should be second nature to the expert chromatographer.

Commercially purchased HPLC solvents seldom have particulates. However, they can contain particles even if these are not visible due to the solvent's refractive index or their size. Such contamination can cause various problems due to dirt accumulating at the head of the column such as:

Changes in V_R
Changes in k'
Decreases in selectivity
Artifacts (spurious peaks)
Drifting with solvent composition changes
Irreversible adsorption and shortening of the column lifetime

Particulates can also lodge in the solvent delivery system, causing pump wear and nonreproducible flow rates. This form of contamination can also

Table 2.2 Effects of Typical Impurities in HPLC Solvents (10)

Contaminant	Possible Source	Effect	Removal
Particulate matter (dust, etc.)	During transfer, unclean vessels	May block in-line filters, lodge in pump seals, or accumulate at column head	Filter through membrane filter
Water	Glassware, solvent preparation or manufacture	Variable column activity, *k′* variation, stability of silicate ester bonded phases	Dry over molecular sieve or anhydrous sodium sulfate
Alcohol	Stabilizer in $CHCl_3$, impurity in hydrocarbons	Similar to water	From hydrocarbons, pass through activated silica; from $CHCl_3$, extract with water, dry with Na_2SO_4
Hydrocarbons (in water)	Organic matter	Baseline instability during gradient elution	Pass through porous polymer column or C_{18} bonded phase
Peroxides (in ethers)	Degradation	Oxidation of bonded phase (e.g., $—NH_2$ to $—NO_2$), reaction with sample, column deactivation or degradation (polystyrene based)	Distill or pass through activated silica gel or alumina
HCl, HBr (halogenated solvents)	Degradation	Column degradation, especially bonded phases, UV absorbence (bromide), stainless steel attack	Pass through activated silica or $CaCO_3$ chips

score the bodies of sample valve rotors, producing crossport leaks which will shorten the valve lifetime (see Section 4.4.5).

The easiest way of removing this form of contamination is to filter the solvent first. Although laboratory filter paper may be adequate, filter paper fibers can pass through with the solvent and later accumulate in filters and at

Table 2.2 (continued)

Contaminant	Possible Source	Effect	Removal
BHT, hydroquinone	Antioxidants in THF	UV-absorbing	Distill
Dissolved oxygen	Solvent preparation	Degrades polystyrene based packings, oxidizes β,β'-oxydipropionitrile, may react with sample	Degas solvent with vacuum or heat
Unknown UV-absorbing	From manufacture	Baseline instability or drift during gradient elution, high detector background	Use activated silica or alumina, or distill for organics, recrystallize or pass over ion-exchange column for inorganics
High boiling compounds	From solvent manufacture	Contaminates collected sample in preparative HPLC	Distill
Algae in water	Growth during prolonged storage	Can plug in-line filters, column entrance frits	Distill from alkaline permanganate or discard

(Printed with permission from the Association of Official Analytical Chemists.)

the head of the column. Commercially available filters, such as those made of mixed cellulose esters having a 0.22 micron pore size for aqueous media and of polytetrafluoroethylene having a 0.2 micron pore size for organic media (Millipore), or HPLC solvent filtration kits (Waters Associates) work fine. A critical factor in filter selection is chemical compatibility. Filter media may contain surfactants, glycerin, and so on, which can wash through with the filtrate or, which may be hydrophobic, preventing filtration of aqueous solutions. Regardless of the filter used, it is recommended to prerinse it first, discarding the first passes of solvent in order to wash the filter clean of any potential soluble contamination. Using a different filter for each solvent is preferred. Any laboratory filtering funnel with a vacuum/aspirator attachment for speeding up filtration is satisfactory. Stainless steel filters, although more expensive than glass, last the longest and perform year after year without surface problems, contamination, or hazardous chipping and cracking from poor handling techniques.

The filtering step, if carried out with an aspirator or a vacuum pump in an evaporating flask, can also serve another key preparatory step, namely, degassing. With all aqueous solvents this is important, and although not always essential with straight organic solvents, it is good CPMA. The lack of degassing, particularly with aqueous systems, can produce bubbles due to degassing at the outlet of the system. Under pressure the solubility of oxygen and nitrogen increases. It seldom causes a problem until elution at the detector, where the pressure dramatically drops to about 100 psi or atmospheric conditions. The dissolved gas leaves the solvent and forms a bubble which will cause erroneous results and havoc with the detector. Methods to handle this are discussed later (see Section 6.6.1). With aqueous systems simple aspiration for 5–20 min, depending on your needs, can be sufficient. Some recommend that in difficult situations the solvent be heated or even the water be first preboiled. If the latter is done, care must be taken to cool the solvent with little head space exposed to the air and return it to room temperature prior to using it as a mobile phase.

2.3.2 Multiple Phase Systems

In the majority of cases the mobile phase is a binary or tertiary system. When preparing these mobile phases, care and clear descriptions of how this is done can be crucial, particularly if the procedure is intended to be routinely and carefully followed by others, and where the values of α are close.

It is good practice to first filter each solvent prior to mixing. Differences in the individual solvents' partial pressures can cause compositional changes when they are filtered as a mixture, and the user may be unaware of the changes in the A/B concentration ratio.

The manner in which solvents are added together can be crucial, and either of the following methods may be employed. No matter which technique is used, it should be noted for other chromatographers to follow.

Using clean 1 liter volumetric glass cylinders, measure the volume of solvent A and add this to the solvent reservoir. Then measure solvent B in a separate 1 liter volumetric glass cylinder and add this to the solvent reservoir; for example, 400 ml of chloroform and 600 ml of *n*-hexane. The total volume should be 1000 ml. This is addition method 1, A + B.

In the second technique you similarly measure solvent A, and bring the volume to 1000 ml with solvent B. This is addition method 2, A + B (q.s.). However, there is a difference between methods 1 and 2, which is particularly noticeable with water-alcohol systems. Every chemist knows that the addition of 50 ml of water to 50 ml of methanol does not produce 100 ml of a 50/50 solution, but rather a solution less than 100 ml. Therefore if A + B (q.s.) is used, the solvent ratio is different from when A + B is used. For HPLC systems, which are critical within a few percent of the water/organic solvent ratio, this difference can be very important. Either system can be used, but consistency in your laboratory, written procedures, and published works will add clarity to your work and allow others to duplicate your systems, as well as helping you in your method development.

Lastly, prepare the mobile phase fresh daily. Usually 1 liter will last 8 hr for most systems. Daily preparation assures that nothing has changed with the mobile phase. Storing the solvent made on Monday for use all week may be satisfactory, but it is not recommended, even if the solvent is not very volatile.

2.4 SOLVENT RESERVOIR

Many different containers from solvent bottles to Erlenmeyer flasks to an internally contained solvent delivery system reservoir are used. Whichever you use, several considerations are important.

1 Use a container with a small covered opening. This prevents or lessens evaporation, which can change the mobile phase A/B ratio (binary system), and prevents changes in k'. Cover the reservoir opening, for example, with aluminum foil. Aluminum foil is lint free and shapes easily to the contour of the container. A hole is easily made through the foil for the solvent line (tube) to the solvent delivery system. With solvent bottles and Teflon delivery tubing, drill out a hole in the bottle cap just under the size of your tubing and pressure fit the tubing through the hole. Be careful not to constrict the tubing because this can cause flow restrictions and possible pump starvation. A few caps such as this can be used time and time again. Finally, this reduces the vapors in the room, which is critical to meeting today's safety standards (see Section 8.2).

2 Be aware of the material from which the container is made. Plastics are

not recommended. Plasticizer leaching, as described earlier with distilled water containers, is a definite problem and will in effect result in an additional yet variable component(s) added to the mobile phase. Glass can also be leached, for example, of metal ions. This is another reason, particularly with aqueous systems and those at pH> 8, why daily mobile phase preparation is important. At extreme pH a nonglass reservoir such as stainless steel would be preferred.

3 Maintain a nitrogen blanket over the mobile phase when using stabilizer-free solvents. It may also be needed with readily oxidizable solvents.

4 Continuously mix the mobile phase to maintain homogeneity. A magnetic stir bar and stirrer are recommended. Different opinions exist as to slow or fast mixing speeds. Fast speeds ensure good mixing. However, this may generate heat with the consequences of compositional changes from evaporation due to the solvent's partial pressure, or increased and variable temperatures in the incoming mobile phase resulting in long term drifting observed at the detector. Slow speeds avoid heat buildup, but may not ensure good mixing. The objective is homogeneity of the mobile phase at 1000 ml as well as at 100 ml. It is not always possible for the final milliters in a solvent reservoir to be representative of the majority of the mobile phase used due to concentration gradients and evaporation which can occur no matter how careful the user is. For this reason, using the last 100–500 ml is not necessarily recommended. Drifting and changes in k' may occur. However, good mixing or large volumes of mobile phase, for example, 2 liters a day when 1 liter is definitely needed, can ensure homogeneity. Each chromatographer will have to adjust his/her system accordingly. Finally, do not place the solvent reservoir directly on the magnetic stirrer. Magnetic stirrers create heat which can affect the mobile phase composition as previously mentioned. A cork ring between the solvent reservoir and the magnetic stirrer works very well, and no negative effect to the stirring of the magnetic bar occurs while adequate space is provided for dissipating any heat from the stirrer.

5 Be careful when positioning the solvent reservoir in the laboratory. This can be very important because of potential temperature gradients affecting the mobile phase. Do not position the reservoir or the HPLC system directly under the ventilation ducts in the ceiling or in the line of drafts from windows, doors, or hoods (see Section 6.3). Most

laboratories are under positive or negative pressure compared to a hallway. In either case drafts through the door can be created. This becomes critical, particularly at weather extremes, creating drafts on the equipment. Furthermore do not allow the reservoir to be in the direct light of either the morning or the afternoon sun. Drifting or changes in drifting will occur throughout the day. This is particularly crucial with refractive index detectors, but it can also occur with UV detectors. Block the windows with aluminum foil, have blinds installed, or move the HPLC system.

6 The tubing from the reservoir can be stainless steel or Teflon. Both are satisfactory. Teflon varies in wall thickness and can allow air through its walls. Usually this is not critical at this point in the HPLC system, since it is under very little pressure (see Section 6.6.1).

7 Keep the reservoir above the solvent delivery system. This assures a good siphon feed to the pump and avoids starving the pumping system.

8 Always use a solvent reservoir filter. These are commercially available (Waters Associates) at about 20–30 microns. They do not impede mobile phase flow to the pump (unless clogged) and can prevent particules from entering the HPLC system. These filters can be cleaned by back flushing with suitable solvents.

9 Always label the reservoir with the composition of the mobile phase, its percent concentration, and the date of preparation. Experienced users find that this prevents mixups and redoing work. If the same solvents will be used daily and many different users share the lab, labeling your solvent preparation equipment may also be helpful.

2.5 CHANGING MOBILE PHASES

Changing solvents is not difficult and is as simple as their preparation. However, be aware of differences in solubility from solvent to solvent and from mobile phase to mobile phase. The following chart (Table 2.3) lists polarities as well as other properties of HPLC solvents which can aid the user in solvent selection and changeover (12). If dramatic changes in the mobile phase are called for, be sure that a solvent with an intermediate polarity is used to purge the system. Immiscible solvent pools can produce very erratic short term noise and are costly in terms of time to remove. Furthermore, this may affect the lifetime of the column if not remembered in time.

Table 2.3 Characteristics of Solvent Used in HPLC

SOLVENT	$E^{o}(Al_2O_3)$[1]	DIELECTRIC CONSTANT	VISCOSITY (cp 20°)	BOILING POINT °C	REFRACTIVE INDEX 25°C	UV CUT (nm)[2]
Fluoroalkanes	-0.25	large			1.25	
n-Heptane	0.00	1.97	0.41	98.4	1.385	200
Hexane	0.00	1.890	0.33	69.0	1.375	210
n-Pentane	0.00	1.844	0.23	36	1.358	210
Isooctane	0.01	1.94	0.53	99.4	1.404	210
Petroleum ether	0.01		0.3	30 -60		210
n-Decane	0.04	1.991	0.92	174	1.412	210
Cyclohexane	0.04	2.023	1.00	81.4	1.427	210
Cyclopentane	0.05	1.965	0.47	49	1.406	210
Diisobutylene	0.06			105	1.411	
i-Pentene	0.08	2.10		30	1.371	
Carbon disulfide	0.15	2.641	0.37	46	1.626	380
Carbon tetrachloride	0.18	2.238	0.97	77	1.466	265
Tetrachloroethylene		2.30	1.82	121	1.631	290
Amyl chloride	0.26		0.43	108	1.413	225
Butyl chloride	0.26	7.39	0.46	78	1.436	220
p-Xylene	0.26	2.27	0.65	138	1.493	290
i-Propyl ether	0.28	3.88	0.37	68	1.368	220
i-Propyl chloride	0.29		0.35	36	1.378	225
Toluene	0.29	2.37	0.59	111	1.496	285
Mesitylene		2.28		165	1.497	290
n-Propyl chloride	0.30		0.35	47	1.389	225
Chlorobenzene	0.30	5.728	0.80	132	1.525	
Benzene	0.32	2.284	0.65	80	1.501	280
Ethyl bromide	0.37		0.40	38	1.424	
Ethyl ether	0.38	4.34	0.23	34.6	1.353	220
Ethyl sulfide	0.38		0.45	92	1.442	290
Chloroform	0.40	4.806	0.57	61	1.443	245
Methylene chloride	0.42	9.08	0.44	40.1	1.424	245
Methyl-*i*-butylketone	0.43			119	1.394	330
Tetrahydrofuran	0.45	7.58		66	1.408	220
Ethylene dichloride	0.49	10.4	0.79	83.7	1.445	230
Methylethylketone	0.51	18.5	0.43	79.6	1.381	330
i-Nitropropane	0.53	25.5		120	1.400	380
Triethylamine		2.42	0.36	89.5	1.399	345
Acetone	0.56	20.7	0.32	56.5	1.359	330
Dioxane	0.56	2.21	1.54	101.5	1.422	220
Ethyl acetate	0.58	6.02	0.45	77.2	1.370	260
Methyl acetate	0.60	6.68	0.37	57.1	1.362	260
Amyl alcohol	0.61	13.9	4.1	138	1.410	210
Dimethyl sulfoxide	0.62	46.2	2.24	189	1.476	
Aniline	0.62	6.89	4.4	184	1.586	
Diethyl amine	0.63		0.38	55.5	1.387	275
Morpholine			7.42	128		285
Nitromethane	0.64	35.9	0.67	101	1.394	380
Acetonitrile	0.65	35.70	0.37	82	1.344	210
Pyridine	0.71	12.30	0.97	115	1.510	305
n-Butanol		17.8	2.95	117	1.347	210
2-Butoxy ethanol	0.74			170.6	1.419	220
i-Propanol	0.82	18.3	2.37	82.3	1.375	210
n-Propanol	0.82	20.1	2.27	97.8	1.383	210
Ethanol	0.88	24.3	1.20	78.5	1.361	210
Methanol	0.95	32.8	0.60	64.6	1.329	210
Acetic anhydride		20.7	0.89	140	1.389	275
Dimethylformamide		37.6		153	1.427	270
Ethylene glycol	1.11	37.7	19.9	197	1.427	210
Trifluoroacetic acid		8.22		72.4	1.283	
Acetic acid	large	6.15	1.28	118.1	1.372	230
Water	larger	78.54	1.00	100.0	1.332	

1. Solvents are listed in order of increasing elutropic value on alumina. The order is essentially the same for silica but the values are different. The data in this table have been taken mainly from *Principles of Adsorption Chromatography*, L.R. Snyder, Dekker, 1968; and *CRC Handbook of Chemistry and Physics*, 44th Ed., Chemical Rubber Publishing Co., Cleveland, 1961.

2. Wavelength at which solvents become opaque in the ultraviolet.

(Reprinted with permission of Instrumentation Specialties Co.)

REFERENCES

1 E. L. Johnson and R. Stevenson, *Basic Liquid Chromatography*, Varian Associates, Palo Alto, CA, 1978.

2 L. R. Synder, *Principles of Adsorption Chromatography*, Marcel Dekker, New York, 1968.

3 *Modern Practice of Liquid Chromatography*, J. J. Kirkland, Ed., Wiley-Interscience, John Wiley & Sons, Inc., New York, 1971.

4 D. H. Rodgers, *Developing an Analytical Method by High Effeciency Liquid Chromatography*, Perkin-Elmer Corporation, Norwalk, CT, 1974.

5 R. W. Yost and R. D. Conlon, *Chromatography Newsletter*, **1**(1) 5–9 (1972).

6 L. R. Synder, *Journal of Chromatography*, **92** 223–230 (1974).

7 D. L. Saunders, *Analytical Chemistry*, **46** (3), 470–473 (1974).

8 N. B. Godfrey, *CHEMTECH*, 359–363 (1972).

9 *Handbook of Chemistry and Physics*, 56th ed., The Chemical Rubber Publishing Company, Cleveland, OH, 1975.

10 R. E. Majors, *Journal of the Association of Official Analytical Chemists*, **60** (1), 186–205 (1977).

11 *Modes of Chromatography: Liquid-Liquid*, Waters Associates, Milford, MA, 1974.

12 *ISCOTABLES*, G. Hofmann, Ed., Instrumentation Specialties Company, Lincoln, NB, 1977.

3
Solvent Delivery Systems

3.1 INTRODUCTION

The HPLC's solvent delivery system, or pump as it is more commonly called, has probably been the component of the most varied design about which more detailed descriptive material has been printed than about any other part of the HPLC system. Very complete descriptions of past and present pumping systems with photographs and schematics detailing the exact manner in which these components operate already exist (1–5). In addition, nearly every short course on HPLC includes a descriptive session on HPLC pumping systems.

The HPLC pump has been developed into a precise analytical component of the chromatographic system. With the evolution of the small particle HPLC columns (5–75 microns), the classical solvent introduction system known as gravity feed was no longer capable of supporting the necessary flow rate through these columns. Even the occasionally used peristaltic pump could not handle these columns adequately, if at all. Both former pumping systems had been used to deliver solvent at a reasonably constant flow rate. However, the new HPLC column needed a solvent delivery system capable of very precise and reproducible flow rates at pressures up to 6000 psi.

The extensive information available on HPLC pumps is a major advantage to the user. The manufacturers have in most cases provided detailed operation manuals, describing setup and operation. They have provided the chromatographer with extensive troubleshooting guides and charts. The depth of these guides varies with the manufacturer. However, the most important step the user can take in learning how to handle and maintain the solvent delivery system is to carefully read the operations manual and troubleshooting guide. This is another aspect of CPMA. So many users tend to unpack a piece of equipment and start working with it immediately before reading the manufacturer's information. This course of action is a major mistake. Many manufacturers offer setup and startup assistance with an equipment purchase. The user should take advantage of this service and learn from the technical representative all he/she can about the pump's care, routine maintenance, troubleshooting indicators, and corrective procedures. The information already available to the user in these manuals will not necessarily be repeated here. The manuals exist for the explicit purpose and benefit of the user with the specific pump from the particular manufacturer. On the other hand guides for pump selection, a brief description of the types of pumps available, and tips on maintenance and care not always necessarily provided will be discussed in the following sections.

3.2 PUMP SELECTION CRITERIA

Selecting an HPLC pump is not much different from selecting any other piece of analytical instrumentation. The usual parameters of capability needed, versatility, accuracy, precision, ruggedness, size, and cost are all to be considered. The user, however, should also select equipment with maintenance and troubleshooting in mind. Find out from the manufacturer how the system works, what routine maintenance procedures are needed and what their costs are, and how close and quickly troubleshooting service will be available to you. As with any purchase, referrals are your best guide. Most manufacturers will give you names of chromatographers using their equipment. Call them and ask about the maintenance, troubleshooting ease, and serviceability of the equipment.

A list of criteria recommended by Wolf is given in Table 3.1 (6). These parameters are listed in decreasing order of importance, with the most important item given first. The user may choose to change the order of these factors to match his/her needs. However, the list will serve to bring attention to many important factors the chromatographer must consider in minimizing problems and maximizing versatility.

Berry and Karger (5) consider the following parameters the most important:

1. Flow constancy (Variation in flow rate will affect detector sensitivity and interfer with quantitation.)
2. Pulseless flow (This is particularly important to minimize detector noise for trace work.)
3. High operating pressure maximum (Low pressure specifications can hinder carrying out fast analytical separations and shorten the lifetime of partially plugged columns.)
4. Wide range of flow rates (This is important primarily for preparative work.)

Table 3.1 Pump Selection Criteria (6)

1	Ease of changing mobile phase, especially when new mobile phase is immiscible with the original
2	Reproducibility of the mobile phase flow rate
3	Minimum baseline pulsation at the detector
4	Ability to operate at high pressures (3000–6000 psi)
5	Ability to operate at low flow rates (0.5–2 ml/min) but with the capability of 0–10 ml/min for analytical separations and small scale preparative work
6	Chemical resistance
7	Safety pressure limit
8	Ease of flow measurement and/or setting the flow rate
9	Adaptability to gradient operation
10	Ability to use small volumes of mobile phase
11	Instant pressure off/working pressure on switching
12	Low dead volume pumping system (recycle may be desirable)
13	Low maintenance schedule and costs
14	Adaptable to constant composition operation (dial in % composition desired)
15	High reservoir capacity and/or ease of adding to the reservoir
16	Minimum evaporation of mobile phase from the reservoir which can cause change of the mobile phase composition
17	Ability to heat the mobile phase reservoir for operation at elevated temperatures

(Reprinted with permission from *Chromatographia*, Vol. 7, T. Wolf, Factors in Choosing a Liquid Chromatograph, Copyright 1974, Pergamon Press, Ltd.)

5 Ease of solvent changeover (When searching for a mobile phase or changing systems, this can facilitate a faster and cleaner changeover.)

Many of these points are quantified in Table 3.2 (4).

Recognizing these criteria and their limitations can aid the user in preventing as well as recognizing problems and saving development and analysis time.

3.3 HPLC PUMPING SYSTEMS

The solvent delivery systems used in HPLC can be divided into two major categories, constant volume or flow and constant pressure. Constant volume pumps are mechanically driven systems, most commonly using screw driven syringes or reciprocating pistons or baffles. Constant pressure pumps are driven or controlled by gas pressure. Various properties of these pumping systems and comparisons between the different types of pumps for specific applications are given in Tables 3.3 (2) and 3.4 (5) respectively. Each system has its own set of advantages and disadvantages with regard to maintenance and troubleshooting. Many of these characteristics are described below.

Table 3.2 Recommended Pump Parameters

Parameter	Mode of Operation		
	Research	Quality Control	Preparative
Pressure, (psi)	5000	3000	1500
Flow			
Range (ml/min)	10	10	20+
Accuracy	±5%	±5%	±10%
Reproducibility	±1%	±1%	±5%
Solvent storage	Unlimited	200–500 ml	Unlimited
Gradient elution capability	Necessary	Not critical	Not critical
Pulse-free delivery	Necessary	Necessary	Not critical

(Reprinted with permission from *American Laboratory*, Vol. 6, No. 10, 1974. Copyright 1974 by International Scientific Communications, Inc.)

Table 3.3 Properties of HPLC Pumping Systems (2)

Pump Type	Advantages	Disadvantages
Mechanical, positive displacement (syringe type)	Delivers high pressures pulse-free Flow rate essentially independent of carrier viscosity and column permeability Rapid pressure buildup Convenient electronic flow	High cost, particularly for a two pump gradient system Limited solvent capacity Inconvenient in changing solvents
Mechanical, reciprocating and diaphram	Essentially a constant carrier flow rate regardless of solvent viscosity and column permeability External solvent reservoir Small internal volume Moderate cost with some models Simple mechanical flow control Continuous operation	Pulsating output which can produce detector noise and limit sensitivity Limited range of volume output Pressure buildup slow
Pneumatic, Simple direct pressure (direct displacement and pressure regulated)	Rugged Inexpensive Easy to operate Pulse-free Pressure buildup fast	Limited mobile phase capacity Limited pressure output Flow rate dependent on mobile phase viscosity and column permeability Inconvenient for solvent changeover Difficult for gradient operation

Table 3.3 *(continued)*

Pump Type	Advantages	Disadvantages
Pneumatic, pneumatic amplifier	High pressure at low cost Relatively pulse-free Convenient flow control Easy to change mobile phase Rapid pressure buildup	Flow rate dependent on mobile phase viscosity and column permeability

(Reprinted with permission of the publisher.)

3.4 MECHANICAL PUMPS

3.4.1 Screw Driven Syringe Pumps

These pumps are driven by an electrical motor displacing the solvent with a syringe at a constant rate. They can attain high pressures up to 7000 psi. The mobile phase is contained in a stainless steel cylinder of either 250 or 500 ml capacity (Figure 3.1).

The pump has a Teflon seal at the end of the piston. Like any liquid seal, it can wear with time, leak, and need replacement. The first consideration is to obtain the correct replacement seal in both size and material. Replacing the seal is not difficult. Retract the piston, remove the solvent reservoir cylinder, and replace the seal. Before assuming that the replacement has been satisfactory, reconnect the solvent reservoir and make several trial runs with the syringe. Disconnect the reservoir and examine the seal. If the Teflon retainer ring and seal have remained on the piston, the replacement has been satisfactory. However, the Teflon retaining ring can dislodge from the plunger and remain jammed inside the solvent reservoir cylinder if it has not been fitted properly or if the wrong size has inadvertently been used. It will usually come off on the up stroke. If this happens *do not* use screwdrivers, rulers, spatulas, etc., to remove the retainer ring and/or seal. First of all it

Table 3.4 Comparisons of Some Common Types of Pumping Systems

Pump	Chromatography				Elution mode		Capacity	
	LSC	LLC	IEC	SEC	EG	IG	Anal.	Prep.
Syringe type	++	++	++	+	−	+/++	++	+
Reciprocating								
Dual-head, special drive with compressibility correction	++	++	++	++	++	+/++	++	+
Single-head, sinusoidal drive	+	+	−	++	+	−	+	+
Multiple-head, sinusoidal drive	+	+	+	++	+	−	+	++
Constant pressure								
Direct gas displacement	+	+/−	−	+	−	−	+	−
Gas amplifier	++	++	+	+	−	+/++	++	++
Pressure regulated (gas)	+	+	−	+	−	−	++	+
Demands								
Flow rate	M	M	L	M			M	H
Pressure	M/H	M/H	M/H	L			M	L/M
Detection limits	M	M	M	H			L	H
Eluent volume	M	M	L	L			M	H

Abbreviations: LSC, liquid-solid chromatography; LLC, liquid-liquid chromatography; IEC, ion-exchange chromatography; SEC, steric exclusion chromatography; ++, optimum in terms of pump performance factors; +, usable in terms of pump performance factors; −, not desirable in terms of pump performance factors; L, low, M, moderate, H, high; EG, external gradient; IG, internal gradient.

(Reprinted with permission from *Analytical Chemistry*, Vol. 45, No. 9, Copyright 1973 by American Chemical Society.

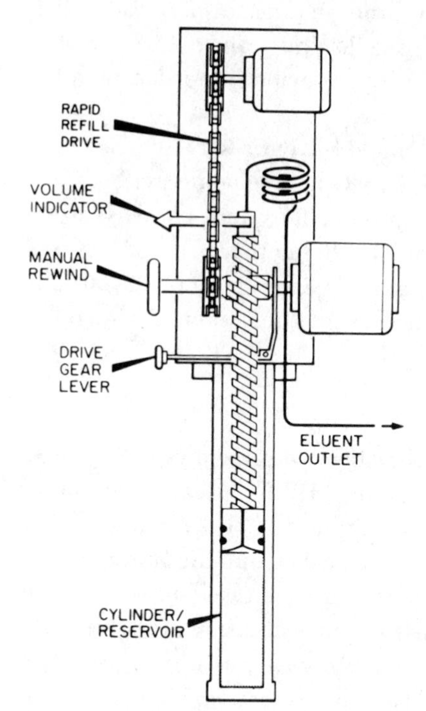

Figure 3.1 Cross-sectional diagram of a screw driven syringe pump (photo courtesy Perkin-Elmer Corporation).

usually is lodged at the far end of the cylinder, and second, and most important, the inside surface of this cylinder wall *must not be scratched.* If it is, it will leak, and it will have to be replaced. These cylinders are made of heavy 316 stainless steel and are very expensive. If any form of probe is used, be sure it is made of soft wood or Teflon. A flexible pickup tool, number 60B265 from Jensen Tools, Inc., 1230 South Priest Drive, Tempe, AZ 85281 works very well. It is 24 inches long and can easily grasp the O-ring or seal and retrieve it. Again, this is a metal tool, and extreme caution must be used not to scratch the cylinder walls.

Care must also be used in the routine removal and reconnection of the solvent reservoir. It is possible to scratch the inside walls during this maneuver. Extreme caution is also needed when handling the pump motor. It weighs about 80 pounds and is in a high position. It may require two people who should be able to handle it easily; the inexperienced user may not be prepared for its excessive weight. This aspect of maintenance and safety is covered in Chapter 8.

This system in general requires very little maintenance. A bad seal is observed by a leak or a disruption and loss of flow rate. Since it has a motor, some routine oiling is needed. Normal care is obviously needed with the common HPLC fittings.

One drawback with this system is its limited solvent reservoir. The reservoir size and flow rate must be considered when developing methods and carrying out routine analyses. Although replacing the mobile phase is easy, it does require shutting down the instrument. One option is to have two of these pumps in tandem, increasing the reservoir size to 500 or 1000 ml. The user should also be careful not to introduce air into the system during refill.

3.4.2 Reciprocating Pumps

These pumps have small dead volume chambers with reciprocating pistons or diaphragms pushing the solvent through the HPLC system. By means of check valves downstream and upstream of the piston or a diaphragm, the mobile phase is drawn in on the up stroke and pushed into the system on the down stroke. With this type of motion the solvent is released into the system in pulses, which can negatively affect the baseline by causing excessive noise and interfer with trace analyses. However, this noise can also be used as a troubleshooting aid, as will be discussed later in this section. There are three basic types of reciprocating pumps used for HPLC: the single piston reciprocating pump, the dual reciprocating piston pump and the reciprocating diaphragm pump.

Single Piston Reciprocating Pump

A single piston reciprocating pump is shown in Figure 3.2 (1). The piston is in contact with the mobile phase on one end and a cam attached to an electric motor on the other. The pistons are usually made of sapphire, borosilicate glass, or chromeplate stainless steel with a mirror finish (4). The nature of these pistons requires special maintenance practices. First, any time the pump head is removed, extreme care is required not to break the delicate piston. These pistons are under extreme pressure by a very powerful spring. Disassembly is not as difficult as assembly. It is advised to have an experienced technical representative demonstrate this first, and possibly have another person assist when assembling it in the future. The manufacturer can also provide instructions on performing this procedure. Second, the piston is in contact with the mobile phase. It is very important that this

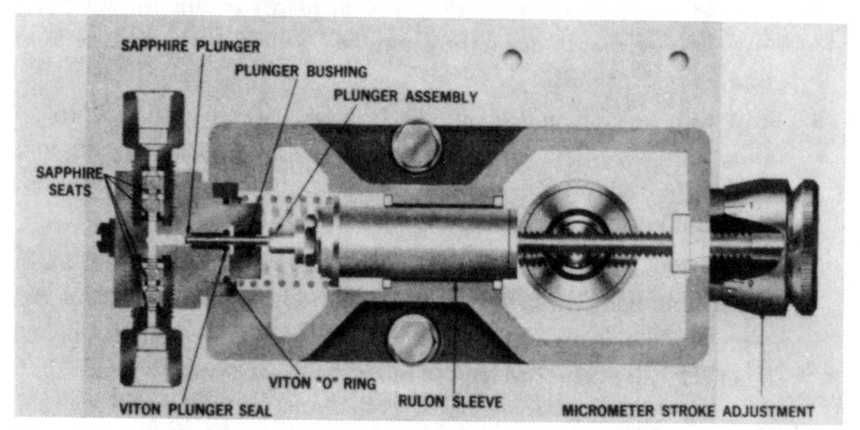

Figure 3.2 Single piston reciprocating pump (reprinted with permission of the publisher).

surface be clean and not scratched. If not, leaks can develop which will seriously affect the flow rate. A leak can be observed by actual mobile phase leaking from the pump head, irregular flow rate, and/or a drop in the operating pressure for the same flow rate as previously used.

One of the biggest problems with these pumps, as well as with any pump having check valves, are dirty, sticking, or malfunctioning check valves. These check valves are stainless steel balls sealing in sapphire seats. The pump in Figure 3.2 shows a double set of inlet and outlet check valves. When they are not functioning, the mobile phase flow will be irregular, or the mobile phase will not be flowing at all. Check valves in this condition can be either replaced or cleaned. The following cleaning procedure works quite well:

1 Prepare several hundred milliliters of 6*N* HNO_3.
2 Mark the direction of flow appropriately on each check valve. This can be done with an engraving tool, appropriately marked tape, or is already marked by the manufacturer.
3 Remove both check valves from the pump head.
4 Rinse about 100 ml of the 6*N* HNO_3 through each check valve with a 50 ml glass syringe having a cannula adapter and an appropriate male Swagelok fitting. Remember that the solvent can pass through the check valve in only one direction. Trying to go the opposite way can only result in an accident with the acid squirting in several directions.

Also, do this over a laboratory sink with plenty of running water for rinsing the acid wash down the drain, and to be available in case of an accident or acid burn.

5. Rinse with several hundred milliliters of tap water to remove all traces of the acid. Tap water is used for this step because it has enough ions present to remove all traces of HNO_3. If any are left behind, nitrosoamines can build up in the pump head.
6. Rinse thoroughly with distilled water.
7. Reassemble both check valves to the pump head.
8. Resume mobile phase flow to determine whether the cleaning opened the check valves. Be sure to use miscible solvents with water at first. This trial should be carried out with the pump disconnected from the HPLC system.
9. If normal flow is resumed, reconnect the pump to the HPLC system. If there is still no flow or very little, repeat the procedure. During step 4 the check valves can be placed in a beaker of 6*N* HNO_3 and sonicated to help dislodge any dirt from the check balls. If this does not work, either replace one or both check valves or call the service representative. (This cleaning procedure can be done with the check valves attached to the pump head, but it will require complete disassembly from the pump.)

The most common problem with a reciprocating pump is gas bubbles lodged in the pumping head. This is caused by the use of solvents that have not been degassed prior to use, particularly with aqueous mobile phases. A solvent with dissolved air in it can dispell this air by the action of the pump, and the air can remain entrapped in the pump head. The other cause of a gas bubble lodging in the pump head is cavitation of the mobile phase. If too volatile a mobile phase is used, such as pentane, the action of the pump can heat up the solvent enough to cause it to volatilize in the pump head, producing a bubble or bubbles which will lodge in the pump head. However, not all gas bubbles will remain in the pump head; some will travel through the HPLC system and interfer with the detector (see Section 6.6). Gas bubble formation can be prevented, or at least drastically reduced, by degassing the solvent first (see Section 2.3.1) or by not using very volatile solvents like pentane.

If gas is trapped in the pump head, the flow rate will not be constant, the baseline will be noisy and/or drift, and the pump may sound louder since it is starved for liquid. The most common way to remove gas bubbles is to redissolve them into a solvent or the mobile phase. The first technique is to put pressure on the pump. Disconnect the pump from the HPLC system and allow it to build up pressure near its pressure limit. This can be done by attaching a cap or plug to the pump outlet, closing tightly down on the cap or plug until the pressure rises close to the pump limit, and then cracking the fitting to relieve the pressure. When the pressure is relieved, let the solvent flow into a beaker or other suitable container. Do this in a laboratory hood or well ventilated area. Repeat this process several times until the gas is dislodged. Remember to use degassed solvent to clear the pump head.

Another technique is to disconnect the mobile phase inlet line to the inlet check valve and the outlet line from the outlet check valve. With the pump on, inject methanol into the pump head with a 1–5 ml glass syringe with a cannula adapter and corresponding fittings. Air bubbles are more easily redissolved in straight methanol under pressure than in water. The user may observe air bubbles being expelled from the outlet check valve. Be sure to do this in a hood or well ventilated area and over a pan or tray to catch the solvent spilling over the outlet check valve. A variation on this technique is to pump straight methanol through the pump at full operating stroke. Do this similarly as with the syringe technique and with the outlet check valve disconnected from the HPLC system. Bubbles may be seen being expelled from the outlet check valve. Be sure to purge the pump with miscible solvents and return to the original, degassed mobile phase. This will avoid possibly ruining a good column and/or considerable system flushing.

The major disadvantage of the reciprocating pump is the baseline noise and the interference with highly sensitive detector studies. It is caused by a pulsing flow rate produced by the reciprocating action of the pump. These pumps are designed for constant volume output, not for constant pressure. The interference with the detector is discussed in Chapter 6. To minimize this effect, pulse dampeners are commonly used. They are usually coils of long lengths of small ID stainless steel tubing, similar to a bellows, which take up the pulsing energy by allowing the damping coil to expand and contract. Sometimes several are used either in parallel or in series with the mobile phase flow. Pulse dampeners can be purchased from Waters Associates, the Milton Roy Company, Glenco Scientific Company, Handy & Harman Tube Co., and Alltech Associates. Other means of reducing this noise is to use a dual head reciprocating pump (see Section 3.4.2) or flow feedback systems. Remember that these devices add dead volume to the HPLC system. Regis Chemical Company sells an electronic active filter (No.

SC-101) to reduce this noise at the detector output. None of these techniques may completely eliminate the pulse effects, but they will significantly reduce them.

Depending on the back pressure of the column, there may not be enough pressure buildup to satisfactorily operate the pump at low flow rates. Many reciprocating pumps do not accurately deliver low flow rates, particularly under no load. One remedy for this is to add solvent restrictors in series with the mobile phase. This can be 10, 20, or more feet of 0.009 inch ID stainless steel tubing, or solvent restrictors may be bought (Waters Associates and the Milton Roy Company). Again, they will add dead volume to the HPLC system.

A newer version of the single piston reciprocating pump, such as the Altex 110-A, utilizes a cam driven piston (Figure 3.3) which takes into account the piston displacement required to compress the mobile phase and pump seals. This concept reduces the pulse noise, minimizes flow rate irregularity due to changes in the HPLC system back pressure, and minimizes the problems of cavitation and small gas bubbles (7).

As with the previously described single piston reciprocating pump, leaks can similarly be detected at the fittings. The piston seal is easily changed by

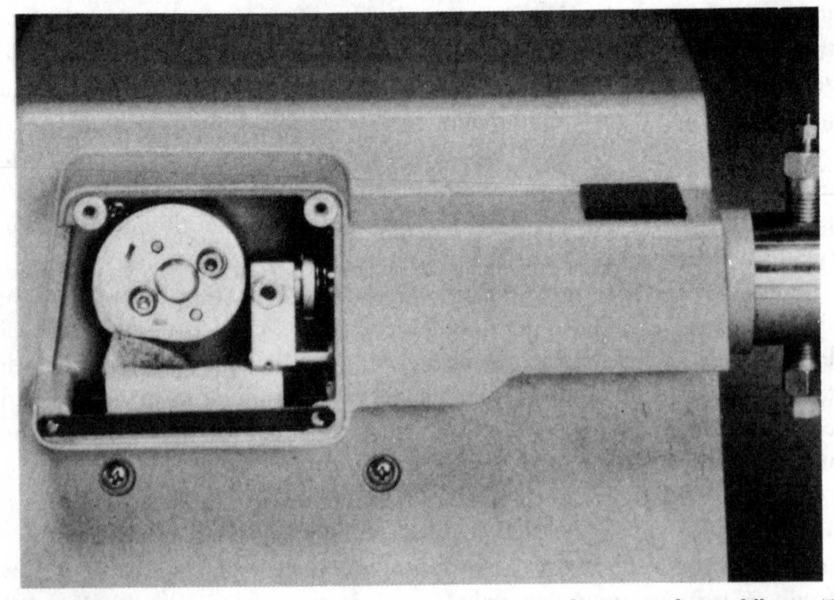

Figure 3.3 Side view of the model 110-A pump showing the cam and cam follower (7) (reprinted with permission from *American Laboratory*, Vol. 10, No. 1, 1978. Copyright 1978 by International Scientific Communications, Inc.).

removing the pump head and using the manufacturer's recommended tool kit. This pump also has a 5μm filter (Figure 3.4) in the outlet check valve to protect the outlet check valve and the HPLC system from particulates (7). If the pump does not have such a built-in filter, a 2–5μm filter can be placed between the pump and the sampling valve. These filters are inexpensive and can protect the HPLC system from solvent particulates as well as particulates from pump wear.

Dual Reciprocating Piston Pump

The dual reciprocating piston pumps were developed to maintain the concept of constant flow but minimize the effects of pulsation. A schematic of a dual reciprocating piston pump is shown in Figure 3.5 (4).

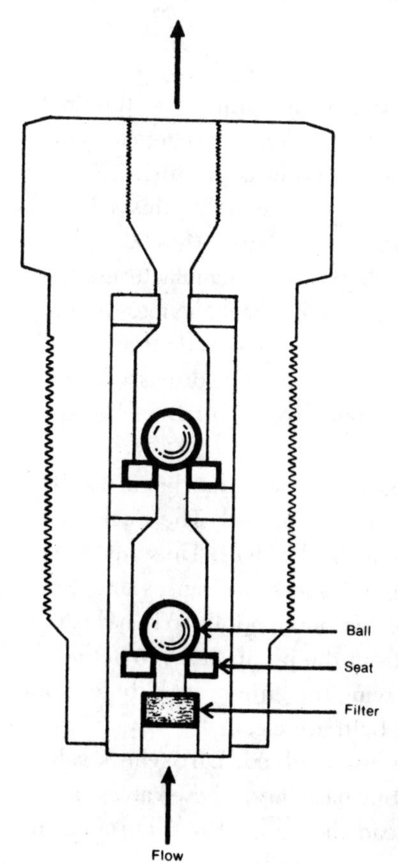

Figure 3.4 Outlet Check Valve of the Model 110 Pump (7) (Reprinted with permission from *American Laboratory*. Vol. 10, No. 1, 1978. Copyright 1978 by International Scientific Communications, Inc.).

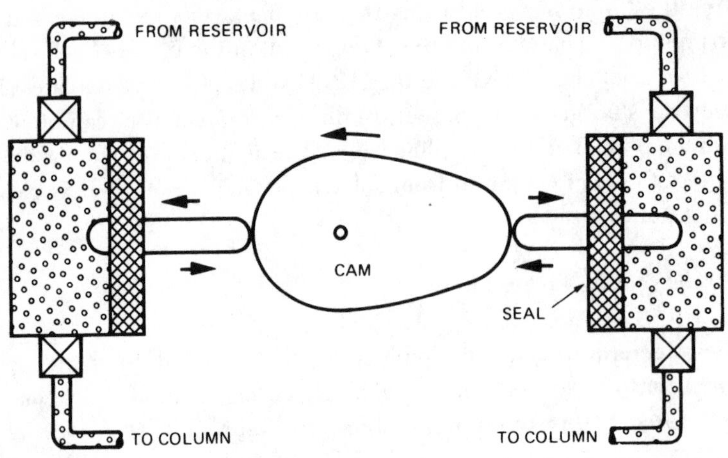

Figure 3.5 Schematic of a dual reciprocating piston pump (reprinted with permission from *American Laboratory*, Vol. 6, No. 10, 1974. Copyright 1974 by International Scientific Communications, Inc.).

The major advantage of the dual reciprocating piston pump over the single piston pump is the reduction in pulsating output. From the schematic alone one can see that as one piston is retracting, the other is pushing. They are synchronized to minimize any pulse lag. As an example, this effect is diagramed in Figure 3.6 with the Waters Associates M-6000 Solvent Delivery System. In addition to this advantage, some manufacturers have added electronic pulse compensation and/or flow feedback devices to better control the output. The user should find that these systems do dramatically reduce pulse noise. If the noise is too great, the dampeners discussed for the single piston pumps can be used, or the service representative should be called.

The problem with entrapped gas bubbles, however, has been doubled with these pumps compared to the single piston pumps. The dual piston pumps have two pump heads and two sets of check valves. Thus any one or both heads can have entrapped gas bubbles. The same causes and techniques as described above for the single piston pump apply. When flushing these check valves, do not remove them from the pump head. Whether or not only one pump head is suspected of having the entrapped bubble, both should be treated similarly to ensure that both are clean.

The problems with leaks and pistons are also doubled. Dirty check valves hinder an accurate flow rate and cause drifting baselines. These valves can be cleaned as described above. The pump head should be carefully removed,

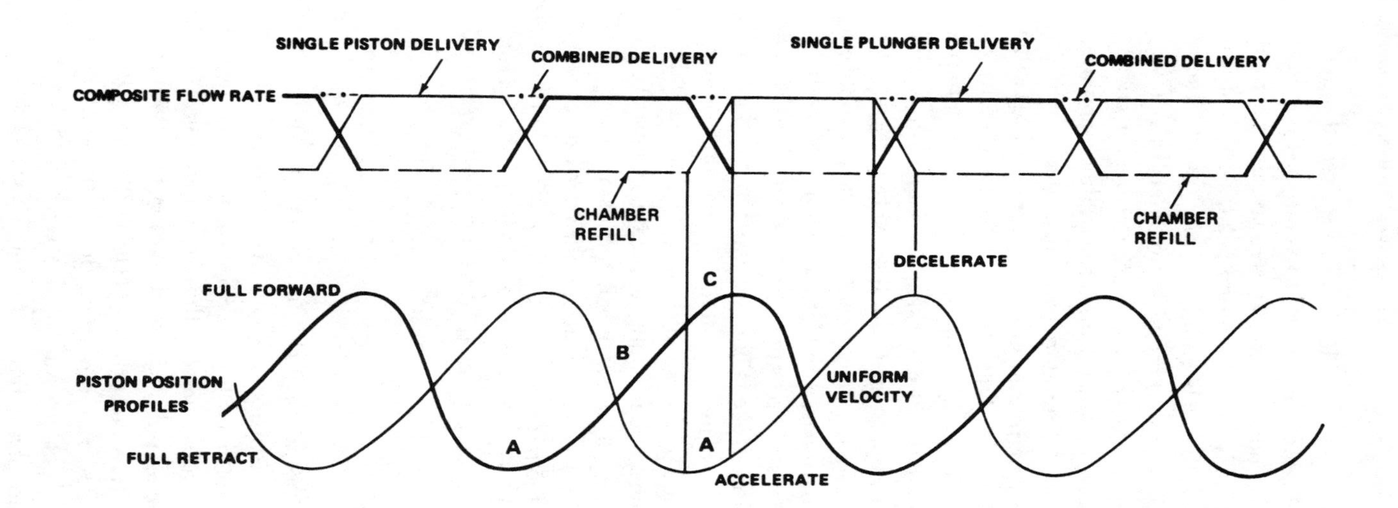

In the Model 6000 Solvent Delivery System, the output from each of two positive displacement chambers is superimposed. This produces a uniform, pulseless, composite flow, achieved by complex piston displacement profiles shown above.

The slope of the piston position curve corresponds to its solvent delivery rate. The forward motion of the piston begins at Section A. In this region the slope increases, resulting in a corresponding gain in flow rate. During Section B, slope and flow rates are constant, with all flow provided by the single piston. The chamber of the other delivery head fills during this interval. In Section C the piston slows down, **and the other piston begins its forward movement, Section A, with the net result of a constant, combined solvent delivery rate.**

Figure 3.6 Dual reciprocating piston pump output flow (courtesy of Waters Associates, Inc.).

but the piston retracted as far back as possible. It can be flushed and/or sonicated with 6N HNO_3 as described for the single piston pump. Both dirty pistons and worn piston seals will produce leaks. These leaks can start out slowly and build to a visible dripping below the pump head. Usually it is a seal, such as one made of Rulan, that has simply worn. Initially gently tightening down on the pump head will stop the leak, but later the seal will have to be replaced. When replacing the seal, check the piston for cracks, surface damage, dirt, and discoloration. Most pistons are made of sapphire, although graphite pistons are also used. Therefore extreme caution is needed when changing plungers and replacing pump heads so as not to break the piston. On pumps with an indicator rod, both the piston and the indicator rod must be centered. The head must be carefully and evenly tightened down, just snug enough and to the point of not leaking. Trial and error will determine when this has been accomplished. If the head is tightened down too far initially, there will be no clearance to tighten it further for small leaks that develop later. If the head is replaced out of alignment, this will be apparent immediately on the initial startup either by a binding indicator rod (which demonstrates that the piston stroke is operating), or by breaking the piston. To replace a bad or broken sapphire piston, it is recommended to have any special tools recommended by the manufacturer. The manufacturer may supply a specification drawing to be used in making this device. When replacing a piston or reassembling the pump head, the user should have spare pistons and indicating rods. Even an experienced user can have an accident, and a broken piston without a spare means immediate downtime. Some sapphire pistons, such as the one on the Altex pump, are thicker and slightly more rugged. Head and piston replacement with these is easier and not quite so delicate.

Reciprocating Diaphragm Pump

The third common type of reciprocating pump is the reciprocating diaphragm pump (Figure 3.7). Here the pistons do not come in contact with the mobile phase and therefore do not need liquid seals which can go bad or have to be cleaned. Because of the attachment to the diaphragm, they can be operated at higher speeds, reducing pulsations even further but not necessarily eliminating them. The pistons are driven mechanically or hydraulically. These pumps suffer from flow rate deviations due to the elasticity of the diaphragm material at high back pressures. This can be detected by carefully monitoring the mobile phase flow rate and checking for flow

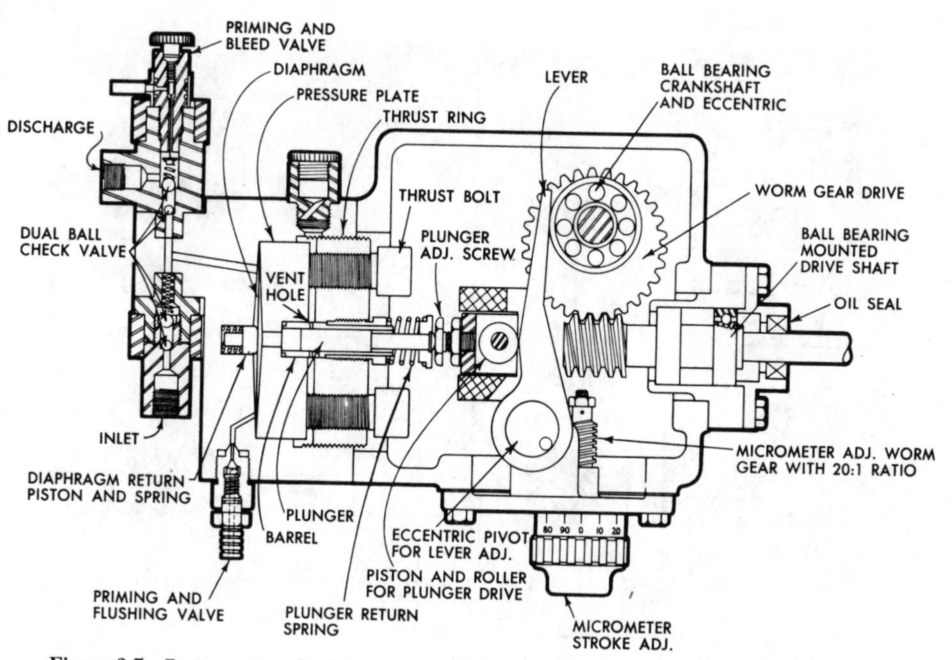

Figure 3.7 Reciprocating diaphragm pump (1) (reprinted with permission of the publisher).

irregularities. A schematic of the Hewlett-Packard 1084 diaphragm pump is shown in Figure 3.8.

A final feature of many of these pumps is an automatic high pressure shut off. It is not uncommon for the chromatographer to be away from the operating system and to return finding that the pump has shut down. First check all electrical conditions, fuses, and the on/off switch in case someone turned the system off. If none of these is the cause, it can very likely be a plugged column inlet or system tubing. A careful trial and error study through each section of the HPLC system can identify the problem area. This shut off can be a valuable troubleshooting aid.

3.5 PNEUMATIC PUMPS

Pneumatic pumps are gas driven constant pressure pumps. The gas pushes either directly on the mobile phase or on a piston or diaphragm which is in contact with the mobile phase. The advantage of these pumps is that they are essentially pulse-free, except when filling the mobile phase reservoir. How-

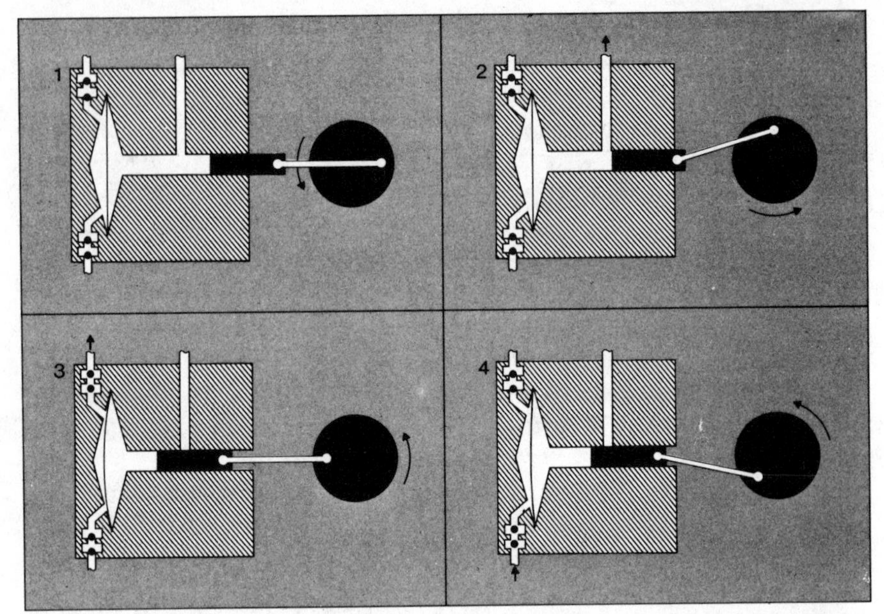

Figure 3.8 Schematic of a high pressure reciprocating diaphragm pump. In diagram 1, the piston is on the end of its fixed stroke. In diagram 2 the piston is moving forward, but no pressure is being developed as displaced oil is pushed into the reservoir. In diagram 3 pressure is being developed as the reservoir inlet has been closed off and, therefore, the diaphragm is pushed, causing the volume of the solvent chamber to be compressed. This, in turn, causes the outlet ball valve to be opened and solvent under pressure to be pumped into the flow system. In diagram 4 the piston is on the back stroke, and the outlet ball valve is closed, while the inlet ball valve is open as solvent is sucked in to replace the solvent pumped out. The piston moves in a fixed stroke, and flow is adjusted by the head being moved forward or backward by a stepping motor. (Courtesy of Hewlett-Packard.).

ever, depending on the type of pneumatic pump employed, they are limited by the gas supply pressure which may not exceed 2500 psi in most versions.

3.5.1 Direct Pressure Pumps

The direct pressure pump uses gas such as nitrogen or helium directly pressurizing the system and forcing the mobile phase through the HPLC system. A potential problem with this system is dissolving the gas into the mobile phase which can produce bubbles in the detector (see Section 6.6). The amount of solvent these pumps can deliver during analysis is limited by the gas supply pressure and the size of the mobile phase reservoir. Both of these points must be considered when selecting these pumps and developing analytical methods. Operating pressures of 4000–6000 psi for microparticulate columns are not typically attainable with these pumps. The user must be thoroughly instructed and knowledgeable in the use of gas cylinders, gas regulators, and gas pressurized systems. This system should have a pressure relief valve for quick shut down. On the other hand these systems are simple to use and inexpensive.

The pressurized coil pump (Figure 3.9) utilizes a small bore stainless steel coil of 200–500 ml volume. As in all direct pressure systems, the mobile phase reservoir is filled externally under atmospheric conditions. The problems of gas dissolution are minimized in this system because the liquid surface exposed to the gas is minimized by the small internal diameter of the holding coil. It is recommended to dispell the liquid near the gas-mobile phase interface where any gas dissolution would have taken place (4). Care must also be taken by the user to properly inspect and handle all valves to avoid accidents.

Another version of the gas displacement pump employs a 500 ml stainless steel cylinder (Figure 3.10). The mobile phase is poured into the stainless steel cylinder, and then the cap with an appropriate gas inlet and mobile phase outlet is secured by hexangonal head screws and O-ring seal. The mobile phase outlet tube has several Teflon baffles about the size of the inside diameter of the cylinder. The baffles minimize the gas-liquid interface and help prevent convection currents.

With this simple direct pressure pump the user must remember that he/she has only a limited volume of mobile phase. Although mobile phase replacement is simple, a complete system shut down is necessary to refill the pump reservoir. The chromatographer *must* be careful to relieve the gas pressure to the pump before disassembly. If the system is still under pressure, the mobile phase will spray all over the user and the surrounding work area, causing potential harm (see Chapter 8). It is therefore essential to have a gas relief valve in the pressurized system.

This pump is also limited in maximum flow rate by the pressure limits of the gas supply. This type of pump typically is operated in the 1000–2000 psi range. Inconsistent flow rates usually will occur when the mobile phase is below 50 ml volume and/or the gas supply is nearly depleted. This is avoided by carefully planning the amount of mobile phase needed to do the desired study and checking the gas supply availability before startup. Many users forget this and have complete analyses or studies ruined by not being able to complete the tests due to insufficient supplies of pressurized gas

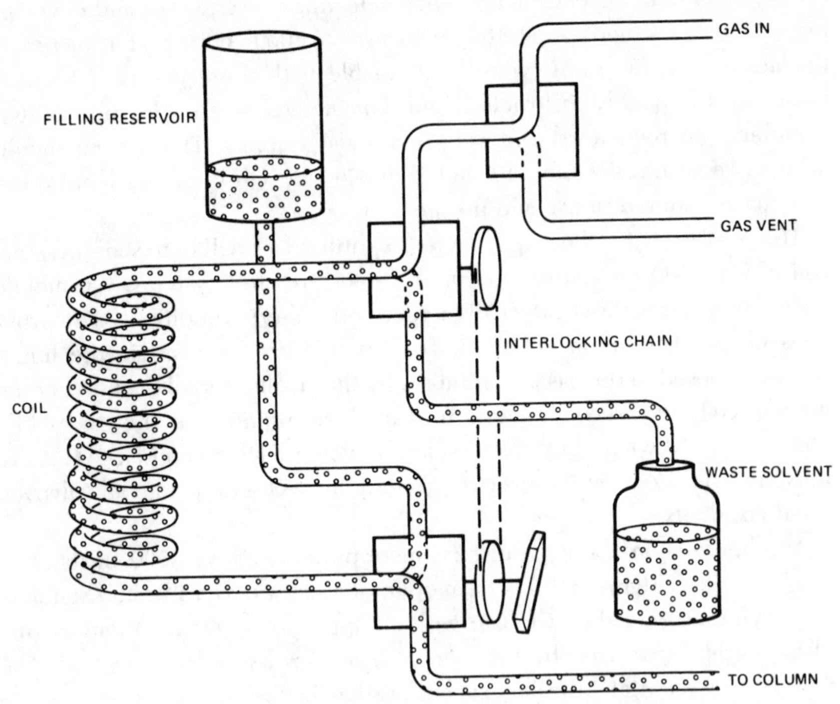

Figure 3.9 Pressurized coil pump (4) (reprinted with permission from *American Laboratory*, Vol. 6, No. 10, 1974. Copyright 1974 by International Scientific Communications, Inc.).

(such as gas cylinders) and/or mobile phase. Good CPMA will aid the efficient chromatographer in avoiding these simple mistakes.

3.5.2 Pneumatic Amplifier Pump

The development of the pneumatic amplifier pump allows users of the constant pressure solvent delivery system to extend their flow rates and operating pressure limits beyond those of the direct pressure pump. This is accomplished by using a piston with approximately a 50:1 surface area ratio between the gas-piston surface and the liquid-piston surface, respectively (Figure 3.11). Pressures in the 3000–5000 psi range are possible with these pumps with gas supply pressures of about 60–100 pounds respectively. Another advantage of this type of pump is its ability to automatically refill after its limited mobile phase supply (about 70 ml) is used up.

These pumps are considered pulse-free and relatively constant in flow

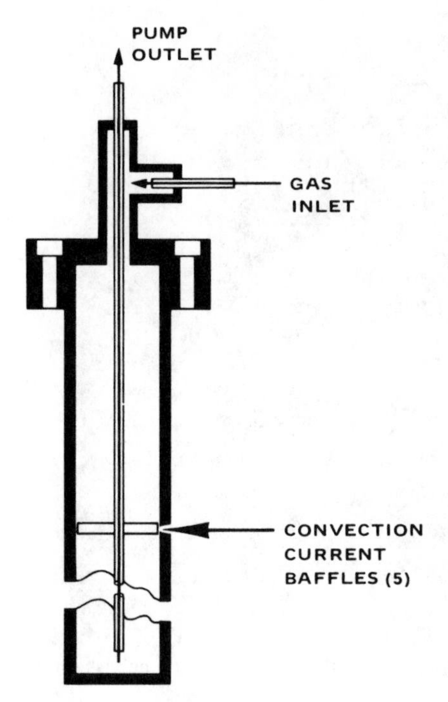

Figure 3.10 Pressurized cylinder pump (8) (courtesy of Varian Instrument Group.)

rate. However, the user must be aware that each time the pump reservoir refills, there is a pressure disturbance in the HPLC system. This disturbance can and will disrupt data output by the detector during trace analysis. A pump refill during elution of a very low percentage component can ruin the data, particularly if manual calculations from a strip chart are required. The less sensitive the detector is set, the smaller the disturbance will be. A good way to determine if the refill action will be a problem is to run the baseline of choice and refill the pump. A disturbance will appear, and the user can determine whether the amount of noise will or will not be a problem. If it cannot be avoided, keep track of the flow volume and avoid elution of sensitive components at that time by carefully staggering injections.

These pumps have relatively few moving parts and are fairly trouble-free. Problems that do occur, such as a worn pump seal, can be detected by flow irregularities and nonreproducible peak heights. If the seal is worn, have a qualified service representative replace it. This is not a simple procedure. Also, the flow rate delivered by these pumps is dependent on any changes in the HPLC system back pressure. A change in the back pressure will cause a corresponding change in the flow rate.

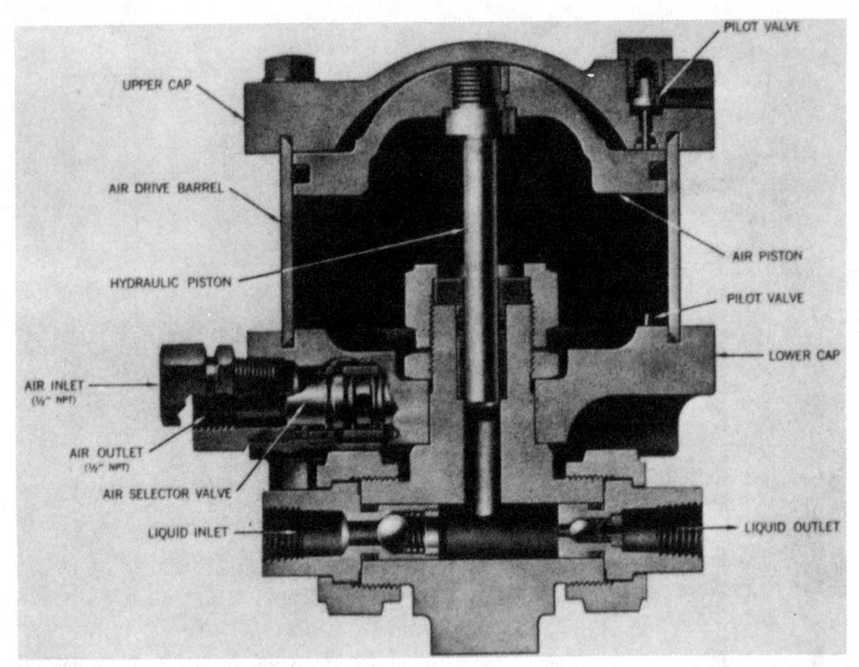

Figure 3.11 Pneumatic amplifier pump (1) (reprinted with permission of the publisher).

Gas bubbles can also get into these pumps. Usually this gas simply limits the amount of liquid the pump reservoir will contain. This can be detected by the pump refilling faster than the flow rate-time schedule would predict. A pump that repeatedly refills is starved and needs solvent. Be sure to keep the main instrument mobile phase reservoir covered during refills. The refilling action can cause the solvent to splash up and out of the reservoir.

Changing solvents with this pump is similar to changing solvents with most pumps. Be sure to rinse thoroughly with intermediate miscible solvents (see Section 2.2.1) if needed, as well as with the next mobile phase of choice. A rapid and trouble-free changeover is carried out by disconnecting the pump outlet from the HPLC system. However, be sure that any other noncolumn system dead volume, such as tubing, valves, and detector cells, is also thoroughly and properly flushed. This is often simplified by replacing the analytical or preparative column with a piece of small ID stainless steel tubing and flushing the entire system.

3.6 PUMP MAINTENANCE CHECK

There are many items that the efficient chromatographer will consider for the care of his/her solvent delivery system, regardless of the types of pumps previously described in this chapter. The following list includes many of these considerations which should become part of the user's CPMA.

1 Determine from the manufacturer and other users of the same equipment what parts need to be replaced, what parts can be easily replaced by yourself, and how frequently these parts need replacement. Once determined, purchase a reasonable supply of each item and stock these in the laboratory. This list will include spare seals, sapphire plungers, indicator rods, O-rings, fittings, cams, pump heads, check valves and associated parts, springs, spring clamps, and so on.

2 Some parts like sapphire plungers do not need frequent routine replacement. However, because this is such a delicate part and can be easily broken during pump head removal for other maintenance work, spare plungers are essential. Also items like springs and clamps which do not usually need replacement can easily be lost during maintenance work. Therefore any parts disassembled during user maintenance work should be stocked.

3 Never store corrosive solvents or buffers in the pump overnight, during weekends, or vacations. Although these parts are made of 316 stainless steel, certain solvents can have corrosive effects on pumps. Spatz has published a list of corrosive chemicals the chromatographer should bear in mind (9). They are listed in Table 3.5. These chemicals can be used, but when the user is finished, the pump and other appropriate parts of the HPLC system should be flushed. In many cases flushing with water followed by methanol works satisfactorily. Passivation of the pump 316 stainless steel has been recommended to minimize the effects of organic acids and chlorides (9). This is accomplished by flushing 20% HNO_3 through the pump (disconnected from the HPLC system) for about 10 min. This is followed by flushing with water until the effluent is neutral to pH paper.

4 Label the pump as to when its seals, plungers, and so on, were last replaced, and keep a pump maintenance log book similar to the col-

Table 3.5 Corrosive Chemicals Used in HPLC (9)

A Potentially corrosive chemicals
  1 Inorganic and organic acids
  2 Inorganic and organic bases
  3 Salt solutions, such as buffer and counter-ion solutions
  4 Complexing agents, such as citric acid, acetic acid, and EDTA
  5 Some organic anhydrides, aldehydes, and sulfur compounds
  6 Various mixtures of individually noncorrosive solvents, such as chlorinated hydrocarbons with isopropanol, acetone, or THF
B Very corrosive chemicals
  1 Strong acid solutions (pH less than 2.0)
  2 Halides in acid solutions

(Reprinted from *Spectra Physics—Chromatography Review*, Vol. 5(2), October 1979.)

umn log (see Section 5.3.4). The label should include the part name, part identification number, and date last changed.

5 When storing a pump for any length of time or in an area with multiple users, label the pump as to the last solvent system used in it, the date, and user identification.

6 Pumps with electric motors may require periodic oiling. Determine what grade oil, if any, is needed, and set up a schedule for routinely checking and filling the oil reservoir (5). Also check for any other lubrication requirements.

7 Always change solvents with miscible solvent systems (see Section 2.2.1).

8 Always degas the mobile phase prior to use to avoid or minimize gas bubbles in the pump chamber (5).

9 Never use very volatile solvents, such as pentane or ether, in pumping systems where the pump action can cause the solvents to volatilize, producing a continuous source of gas bubbles in the system. If this is necessary, be prepared for problems.

10 Upon receiving a new pump, have the manufacturer's service representative set it up and go through all its operations and normal maintenance and troubleshooting procedures. If you have a mechanical and/or electrical maintenance department for servicing your

analytical instruments, let them review the pump schematics. They may help identify the components most likely to go bad.

11 Operate motor driven pumps in well ventilated areas to avoid overheating.

12 Always check that the pressure limiting switch, if available, is properly set.

13 Pump seals, material, tubing, and so on, can undergo dimensional changes under HPLC operating pressure. This can affect the delivered volume by as much as 10–20% (10).

14 Liquids can undergo up to 4% compression at HPLC operating pressures. This is a problem during gradient elution with solvents of greatly different viscosities (10).

15 Check the pump heads and fittings daily for leaks by either visual or physical examination (see Section 5.3.3).

16 Check the flow rate of the pumping system periodically if it is a constant volume pump or as needed with a constant pressure pump to be sure that the required flow rate has been achieved. This is easily accomplished with a 10 ml graduate cylinder and a stop watch.

REFERENCES

1 R. Henry, *Apparatus in Modern Practice of Liquid Chromatography*, J. J. Kirkland, Ed., Wiley-Interscience, John Wiley & Sons, Inc., New York, 1971.

2 L. R. Snyder and J. J. Kirkland, *Introduction to Modern Liquid Chromatography*, John Wiley & Sons, Inc., New York, 1974, Chapter 4.

3 J. Q. Walker, M. T. Jackson, Jr., and J. B. Maynard, *Chromatographic Systems Maintenance and Troubleshooting*, Academic Press, Inc., New York, 1977, Chapter 2.

4 M. T. Jackson and R. A. Henry, *American Laboratory*, **6**(10) 41–51 (1974).

5 L. Berry and B. L. Karger, *Analytical Chemistry*, **45**(9) 819A–828A (1973).

6 T. Wolf, *Chromatographia*, **7**(1) 33–35 (1974).

7 R. Stevenson, R. Henry, H. Magnussen, and P. Mansfield, *American Laboratory*, **10**(1) 41–48 (1978).

8 N. Hadden, F. Baumann, F. MacDonald, M. Munk, R. Stevenson, D. Gere, and F. Zamaroni, *Basic Liquid Chromatography*, Varian Aerograph, Walnut Creek, CA, 1971.

9 R. Spatz, *Spectra-Physics Chromatography Review*, **5**(2) 6 (1979).

10 R. Henry, Advantages of a Constant Pressure Pumping System for High Speed Liquid Chromatography, Liquid Chromatography Technical Bulletin No. 73-1, DuPont Instruments, Wilmington, DE.

4 Sample Preparation and Introduction

4.1 INTRODUCTION

A major advantage of HPLC is its widespread application to the majority of organic compounds as well as to many inorganic compounds. Gas chromatography (GC), which had become so popular in the 1960s and is still highly used today, is dependent on the volatility of the compound of interest. Its versatility has been expanded with the use of derivatization. But in general GC is only applicable to about 20% of the known organic compounds. As opposed to GC, which utilizes either gas-liquid or gas-solid modes of separation, HPLC has four branches of separation: liquid-liquid, liquid-solid, ion-exchange, and size exclusion. With these modes of separation available, the analyst can handle nearly any sample submitted.

However, as in most analytical problems, samples are usually not submitted in simple known matrices. Even with known systems there are usually one or more interfering species present. This is where chemical know-how and the knowledge of chromatographic theory are important to the chromatographer in order to choose the proper mode of separation. Approaches to selecting the separation mode are discussed elsewhere (1–4).

Therefore for the purpose of this chapter, assume that the mode or sequential modes of separation have been selected.

4.2 SAMPLE PREPARATION

4.2.1 Solid Samples

Solid samples have to be dissolved before being introduced into the HPLC system. The choice of solvent is critical. Ideally the analyst should dissolve the sample in the mobile phase. This has several advantages. First, it minimizes the sample solvent peak at the void volume. This is particularly critical with unknown samples where an impurity or peak of interest can be masked by the sample solvent peak. Second, it avoids precipitation on the column. Sample dissolution in a quickly known solvent may speed up sample preparation, but can later cause endless hours of troubleshooting and misinterpreted data due to the sample or a component of the sample precipitating on the column. This occurs when the sample slug dissolves and diffuses in the mobile phase. As the sample solute is dispersed, those components not readily soluble in the mobile phase can precipitate out. If the component of interest does not precipitate, interference from a precipitate may not be noticed until subsequent injections. For example, in a binary reverse phase system of water-methanol it is common to dissolve the sample in straight methanol for ease of preparation or when the sample will not dissolve directly in the mobile phase. If precipitation takes place before or on the column, it is possible to observe unknown and randomly eluting peaks with later injections. These peaks do not maintain the same k' and can elute on or partially cover the peak or peaks of interest. This can occur for two reasons. The precipitated material redissolves in the methanol slug from the next injection and is carried partially or totally off the column. The same effect can happen, but more slowly, with continuous washings of mobile phase. Precipitation can also occur at the head of the column, clogging the inlet, increasing the column back pressure, and restricting the flow. In addition split peaks can be produced (see Section 5.4.2).

All of these symptoms will hinder quantitative analysis. For quick qualitative studies it is possible to use the quick method of preparation. However, these potential problems must be kept in mind. If the sample preparation is done in a solvent different from the mobile phase, column cleaning with a suitable solvent followed by reconditioning with the mobile phase is recommended. This can be done with repeated 1 ml injections of a good solvent, for example, straight methanol, or running methanol through the HPLC system via the solvent pump. This must be followed by flushing and reconditioning with the mobile phase, preferrably overnight. No matter what solvent is used, be sure it is compatible with the column.

Samples which do not readily dissolve in the mobile phase can be coaxed to or can be dissolved in a solvent system that will approximate it. For binary or tertiary mobile phase systems, dissolve the sample first in the solvent component that will dissolve the sample most readily. Follow this by adding or making to volume with the remaining solvent component(s), while keeping the ratio the same or as close as possible to the mobile phase. This may take time, but the few additional minutes can be well worth it. The use of ultrasonics can also aid in sample dissolution, but caution is needed to avoid heat buildup and possible thermal degradation of the sample.

Finally, preparing the sample in the mobile phase maintains system compatibility. As mentioned in Chapter 2, after several hours of use the mobile phase ratio can change. Continual sample makeup with the mobile phase as is can assure little difference in the void volume peak. This can be important when the peak of interest has a low k' as compared to the void volume.

These guidelines are particularly important in the liquid-solid and liquid-liquid modes. The same general rules can apply to ion-exchange or exclusion, but they are usually not as critical.

4.2.2 Liquid Samples

Liquid samples provide the chromatographer with the option of injecting the sample directly. However, the solvent may or may not be compatible with the chromatographic system of choice.

If the solvent is compatible, for example, a water sample and a reverse phase, ion-exchange, or aqueous exclusion system is to be used, direct introduction of the sample can be made. System modifications may be needed, depending on the complexity, interferences, and concentration levels of the sample.

For samples not in a desired solvent or not concentrated enough, two simple procedures can be used. One is to evaporate the sample to dryness and reconstitute the sample with the mobile phase or a more suitable solvent. Guidelines for solid samples discussed in Section 4.2.1 will apply. Partial evaporation can also be used to increase the concentration of a sample, assuming that the sample solvent is compatible with the chromatographic system.

4.2.3 Sample Filtration

It is just as important, if not more so, to filter the sample prior to injection as it is to filter the mobile phase. The choice to filter a sample will be determined by the nature of the sample, its solubility, and interferences or contaminates.

As opposed to the mobile phase which may have a solvent reservoir filter and in line pump filter(s) before the sample inlet and column, the sample's only filter may be the column itself. Whether it is insoluble material or lint from inside the sample container left after cleaning and drying the glassware, insoluble contamination buildup at the head of the column can lead to restriction of the mobile phase flow, increasing the column back pressure, decreasing the column's efficiency, and producing split peaks (see Section 5.4.2).

The chromatographer can use classical filtration techniques, specialized equipment such as a syringe and a 5 micron filter pad in a Swinny adapter (Millipore Corporation), or a commercial sample clarification kit as from Waters Associates. With any filter system it is advisable to determine whether the solute is absorbed by the filter media, particularly for quantitative analyses. This can be accomplished by spiking experiments with known samples of predetermined concentration. Failure to determine this can cause nonreproducible results.

4.2.4 Solvent Degassing

It is advised to prepare samples with degassed solvents. This will reduce the possibility of degassing occurring in the detector cell due to the sample solvent.

For quantitative analysis the sample solution should not be degassed. Degassing causes solvent evaporation, which will change the sample concentration. Degas the solvent before preparing the sample.

4.2.5 Internal Standards

The use of an internal standard is preferred by many analysts for quantitative analysis. Its purpose is to minimize system and procedure variations, thus eliminating variations in precision as a function of sample size.

For the proper use of an internal standard, several requirements should be met (1,4):

1 It should be completely resolved.
2 It should not elute on or over another component, yet have a k' as close to the peak of interest as possible.
3 It should have similar chemical properties to eliminate or reduce differences in detector response between itself and the component of interest.
4 It should be prepared at the same concentration level as the sample.
5 It should be of good purity to prevent adding contamination and spurious peaks to the chromatogram.
6 It should be chemically inert.

4.3 SYRINGE INJECTION

Samples are typically introduced by syringe injection into the mobile phase stream or on-column and by the use of sampling valves (see Section 4.4). Each method is relatively simple to use, but certain problems must be avoided to carry on trouble-free work. The key to sample introduction is precise and reproducible injections. This is especially important with quantitative analyses where the reproducibility of the peak response is dependent upon the precision of the sample introduction.

Direct syringe injection techniques were the first popular method of sample introduction. As HPLC instrumentation evolved, many basic techniques from GC were copied, and sample introduction was no exception. Figure 4.1 shows a diagram of a typical injection port.

Syringe injections through a septum into the mobile phase stream worked particularly well, since many of the early instruments were not operated at pressures much greater than 1000 psi. Syringe injection in HPLC is pressure limited and is not useful over operating pressures greater that 1000–1500 psi. The actual pressure limit will depend upon the type of septum, retaining nut, syringe needle, and so on. Consequently in recent years this method has seen limited use with the development of microparticulate columns and higher operating pressures up to 6000 psi.

The principal problem at higher operating pressures is leakage, which may occur in two ways.

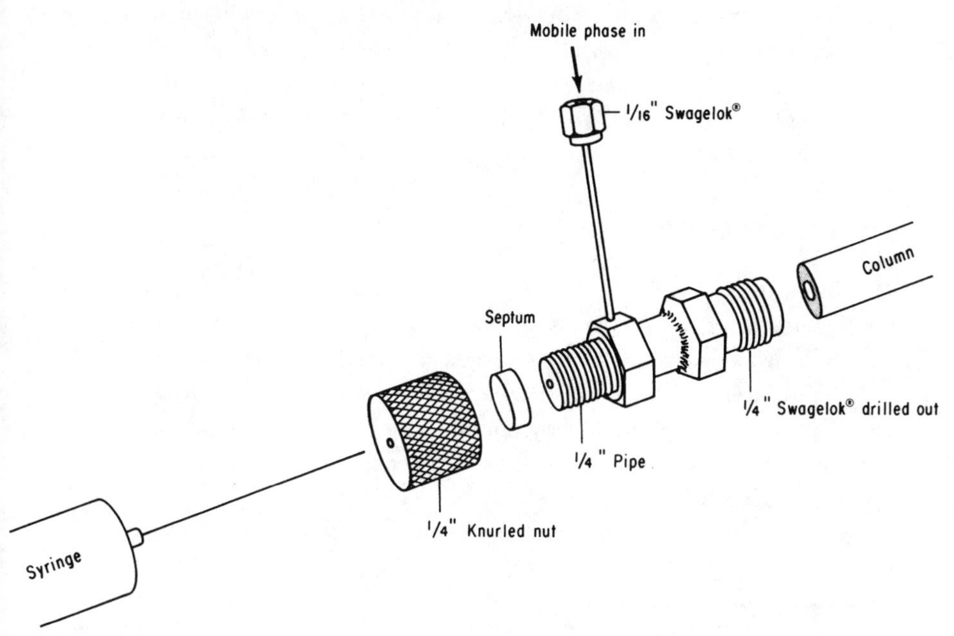

Figure 4.1 Diagram of a typical on-column injection port (reprinted with permission of the publisher).

1 The septum is typically held in place by a retaining cap which, as it is tightened, seals the septum up against the injector assembly. Such a seal can only hold a limited liquid pressure. Leakage will develop between the septum and the surface against which it is forced. The leaks can start very small and may be evident only by the cool feel of the metal retaining cap. In this case the leak is too small to be observed, but evaporation around the cap cools it to the touch. A leak can also burst around the retaining cap, and the mobile phase will run down the face of the instrument. Turning the cap tighter and tighter to make the seal will only distort the septum and increase the possibility of leakage. It will also cause too much pressure on the septum, making it very difficult for the syringe needle to get through.

2 A leak can develop through the hole made by the syringe needle in the septum. This can occur even with a good needle and septum at high enough operating pressures. In order to inject, the syringe needle must pierce and bore a hole into and through the septum. Depending on the condition and angle of the syringe needle tip, this hole or tear

can act only as a limited pressure seal. A leak here will range from a drip to a solid stream of mobile phase shooting straight out, several feet from the septum. This not only prevents operation of the instrument, but it presents a safety hazard, since it can hit the analyst on the body or face.

4.3.1 Choice of Septum

The dimensions of septa for HPLC vary depending on the size and shape of the injector. In general they all resemble the typical GC septum. The septa are made from materials such as standard silicone rubber, perfluorelastomers, and Neoprene®. Septa can be purchased directly from the injector manufacturer or supplier, or they can be made by buying a standard sheet of septum material and cutting out the size needed with a cork borer or other suitable cutting tool.

Several points are important in the choice of a septum.

Septum-Solvent Compatibility

The major problems with the solvents is leaching of the septum material. This will cause baseline drifting, physical and chemical contamination of the HPLC system, a decrease in the detector sensitivity, a drop in the detector linear response range, and a decrease in the septum lifetime. Some commercial septa and their respective solvent compatibilities are listed in Table 4.1.

Septum Hardness

Some septa are made from harder material than others. The softer the septa, the easier it is to make the liquid seal. For example, Viton-A (Table 4.1) is a harder material than BUNA-N, EPR, and WSR. This factor can also determine the injection lifetime of the septum. A harder material cannot withstand as many injections as a softer material.

One method of overcoming a solvent compatibility problem or extending the injection lifetime of a hard septum is to use a "septum sandwich." In this technique two different septa are placed back to back. The harder or solvent compatible septum faces the mobile phase, while a softer and possibly non-compatible solvent septum material faces outside toward the analyst. This allows a harder, solvent compatible material to be in contact with the mobile phase, while a softer material can be used to ensure a better seal and more

Table 4.1 Solvent Compatibility with Some HPLC Solvents

Septum Material	Compatible Solvents	Septum Supplier
BUNA-N	Water, DMF, *n*-hexane	Waters Associates (5)
EPR	Water, MEK, ether ligroin	Waters Associates (5)
VITON-A	Water, Toluene	Waters Associates (5)
WSR	Aqueous solvents, alcohols, THF, most organic solvents	Waters Associates (5)
Silicone rubber	Aqueous solvents, alcohol, polar solvents	DuPont (6)
Perfluorelastomer	Hydrocarbons, nonpolar solvents	DuPont (6)

injections. Most septa will last 10–20 injections. The sandwich technique can increase this by 50% for hard materials.

4.3.2 Choice and Use of the Syringe

Many manufacturers and suppliers distribute HPLC syringes such as Hamilton, Precision Scientific, Scientific Glass Engineering Pty. Ltd. (S.G.E. International), and Unimetrics Corporation. The primary characteristic of these syringes is their ability to be used at HPLC operating pressures (1000–1500 psi). This is accomplished with Teflon plugs around the plunger in the syringe barrel or a precision fitting plunger in the syringe barrel.

In the latter case it is preferred to use a Chaney adapter with this type of syringe (for example, Hamilton) to prevent syringe "blow back." This is a limiting factor for using syringe injection over 1000 psi. Syringe "blow back" is the effect of the mobile phase pressure retarding the injection and pushing the syringe plunger and mobile phase out the back of the syringe barrel. A Chaney adapter can prevent the syringe plunger from leaving the barrel. If the syringe does not have a Chaney adapter, syringe injections at higher allowable pressures can be made safely by holding the plunger and barrel with one hand and guiding the needle through the septum with the other.

The plunger can thus be pressed carefully against the barrel, preventing its movement while injecting the sample.

The syringe should have a fixed needle. This will prevent leaks. However, as in GC, a major problem with syringe users is that they bend and damage the syringe needle. This is done by missing the septum during injections, not easing the needle into and through the septum gently, or by other mishandling practices. If the chromatographer has trouble making injections without damaging the needle, a syringe with a barrel guide (S.G.E. International) can be used. Similarly, the user must be careful not to bend or damage the plunger.

The end of the syringe needle must be free of barbs. Barbs will tear the septum, and septum particles can clog the syringe needle or deposit into the HPLC plumbing, causing flow restrictions between the injector and the column or at the head of the column. Tearing the septum will also reduce the septum injection lifetime and increase the possibility of leaks. The chromatographer should routinely inspect the syringe and keep a supply of spares on hand in the various sizes of choice.

Whenever injecting a sample, it is imperative that no air remain in the syringe nor be injected into the HPLC system. This is easily avoided by overfilling the syringe, inverting it with the needle pointed upward toward the ceiling, and slowly pushing the air out with the excess liquid. Depending on the difficulty of removing an air bubble lodged on the inside surface of the barrel, tapping the syringe or repeated refillings may be needed to dislodge and fill the syringe air-free. Most syringes are made of glass, and any air present in the barrel can be seen. Air in the needle is usually dislodged easily by dispelling excess sample.

A common user problem is a clogged or plugged syringe. This occurs either from septum particles lodging in the syringe needle, or from sample drying out in the syringe. Both can be avoided by routinely using sound syringe handling practices. The former can be avoided by using a barb-free needle with a 20° angle. The latter is avoidable by always rinsing the syringe with a clean suitable solvent and air drying it after every use and before storage.

In the event of a clogged syringe, several cleaning techniques can be used. For needle clogs a fine wire may be used to dislodge or ream it out. In the case of septum particles lodged in the needle, care must be taken not to force the particle(s) further up into the needle and barrel. Another procedure is to carefully fill the syringe barrel with a suitable solvent from the back end, insert the plunger, and try to push the solvent out the needle. Be careful of pressure induced "blow back" and of cracking the syringe barrel by exerting too much pressure.

For dried out samples, reaming with a fine wire or adding a suitable solvent through the back of the syringe barrel and proceding as described above can be used. The syringe can also be soaked in a suitable solvent with and without ultrasonics.

4.3.3 Stop Flow Injections

Syringe injections become very difficult above 1200–1500 psi. It becomes difficult to hold the plunger from blowout manually, and some syringes cannot withstand these operating conditions. However, the chromatographer can use the syringe/septum method by employing the technique of stop flow injections.

Stop flow injections are made by turning off the chromatograph pump or depressurizing the injection port, letting the pressure drop to zero, making the injection, and resuming normal operation.

This technique has not found widespread usage due to the ease and availability of high pressure sampling valves. However, if no valve is available, it is an alternative sampling approach. When using this technique, the following helpful hints will reduce problems.

1. Best success is achieved with constant pressure solvent delivery systems. After the injection is made, the system operating pressure can be resumed quickly, and a usable baseline will recover quickly. When using a constant volume pump, the drifting baseline produced during startup can present a problem for components with low k'.
2. For quantitative analysis use an internal standard. This will minimize the effect of system variations on the injection precision.
3. Maintain a minimum dead volume. This is important to minimize the effects of diffusion from injecting the sample into a static mobile phase.

4.4 VALVES

The use of high pressure sampling valves for HPLC have become the most popular means of sample introduction. This has resulted from the popular use of microparticle columns and solvent delivery systems operating most frequently in the 2000–6000 psi range. A 6-port rotary valve is shown in

Figure 4.2. Although relatively simple devices, valves are frequently a constant source of problems to the chromatographer. These problems can be avoided by following some very simple guidelines.

4.4.1 Operating Pressure Specifications

Know the operating pressure limit for the valve in use. The majority of valves used are rated for operating limits of 1500, 3000, 6000, and 7000 psi. If you have valves whose pressure limits are in question, contact the manufacturer or supplier for this information. The limits are relatively accurate, and matching or exceeding these specifications will definitely cause valve leakage around the rotor. This will result in excessive downtime dismantling the valve, drying it out, and rebuilding.

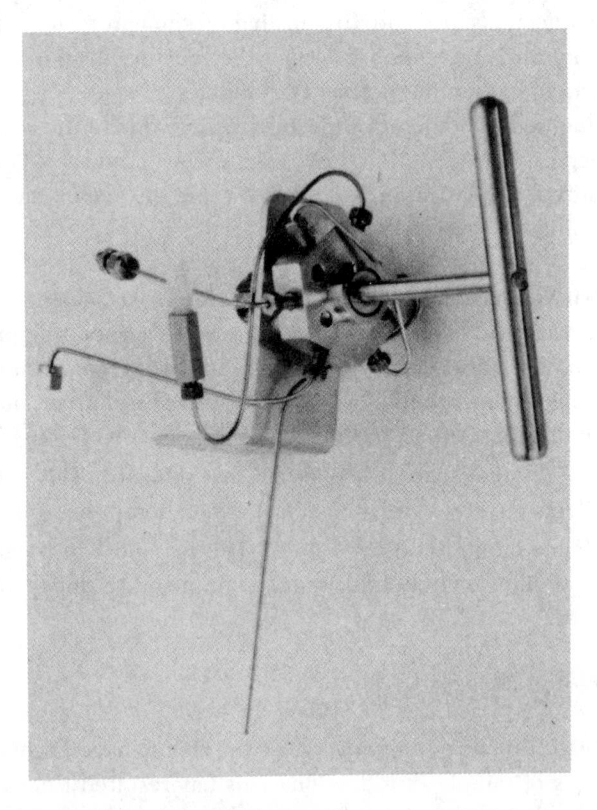

Figure 4.2 A 6-port rotary valve (Altex High Pressure Liquid Chromatography, Altex Scientific, Inc., Berkeley, CA, 1975).

4.4.2 Spare Parts

Have plenty of dead volume fittings, ferrules, and rotors for valve repairs. It is common to replace the fittings and ferrules because of damage and changes made in the system, such as adding new loops or changing the valve's inlet/outlet tubing. Not all fittings and ferrules typically used in HPLC laboratories are interchangeable or useful with these valves. For the best nonleak fittings, use only parts supplied by the valve manufacturer or supplier.

Many valve manufacturers supply spare rotors. The rotor is the seal-handle assembly which is pressure fitted into the stainless steel valve body. The most common damage to valves are crossport scratches on the rotor body, producing leaks. This occurs because of particulates getting between the rotor and the stainless steel valve body and scratching the rotor body by producing grooves between the valve port openings. When replacing a rotor, be sure to use the correct rotor for the valve. Depending on the pressure limitation of the valve, rotors are made from different materials and do not have the same pressure fit configurations to be interchangeable. This is true even among different valves from the same manufacturer. When changing valve rotors, be sure to follow carefully the manufacturer's recommended procedure. The Valco valves, for example, use spring loaded washers to seal the rotor. Depending on the pressure limit for the valve, either all the washers will be oriented in one direction, or half of the washers will be oriented in the opposite direction from the other half.

It is also advisable to have several spare valves. Depending upon usage and damage rate, in-laboratory repairs may not be possible. The damaged valve can be repaired by returning it to the manufacturer. While in repair, the laboratory should have at least one valve for use and still another back up in case of problems which can occur while the one valve is out for repairs.

For laboratories with many HPLC instruments, users, and valves, it is advisable to monitor the repair and location of all the valves. This can be done by carefully engraving a code on the outside of the stainless steel body of the valves. The code should be sequential to accommodate the purchase of more valves. This information can be monitored by use of an HPLC inventory log book. Figures 4.3 and 4.4 illustrate examples of two forms that may be used for this purpose. The HPLC valve inventory log sheet (Figure 4.3) is used to code the valves with pertinent historical information as they are received. The individual valve log sheet (Figure 4.4) is used to monitor the location, repairs, and overall use of the valves.

Date Received	P. O. Number	Manufacturer	Purchased from	Valve Identification	Valve Type	Comments

Figure 4.3 HPLC valve inventory log.

Valve Identification ________________

Valve Location			Repairs				Number of Injections	Injection Technique	Comments
Date	Lab	Current Use	Repair Company	Date Sent	Date Returned	Repair Number(s)			

Figure 4.4 HPLC individual valve log.

4.4.3 Valve Fittings

Leakage is commonly caused by mishandling the low dead volume valve fittings. These fittings, like all HPLC stainless steel fittings, press a cone shaped ferrule onto a stainless steel tube. This is tightened down into the valve body (Figure 4.5).

When tightening the ferrule and fitting in place, the novice frequently applies too much torque to make the seal. These fittings, like all HPLC equipment, *must* be handled gently and a "herculean" effort on a nut will not seal but only tend to distort the ferrule and cut a groove in the surface of the tubing. These fittings must be tightened very carefully.

When sealing stainless steel tubing to the valve body, carefully butt the tube up against the inside surface of the valve body port. With the tube snuggly in place, slide the ferrule over the tubing into the port and carefully tighten it down with the nut. *Do not overtighten!* Stop once a snug fit is made. It is better to run the system under pressure and retighten the nut than to overtighten. Using the correct size wrench will help. The wrong wrench, such as a large crescent wrench, can easily provide too great a torque due to its longer handle, making it easy to overtighten and gall the fitting.

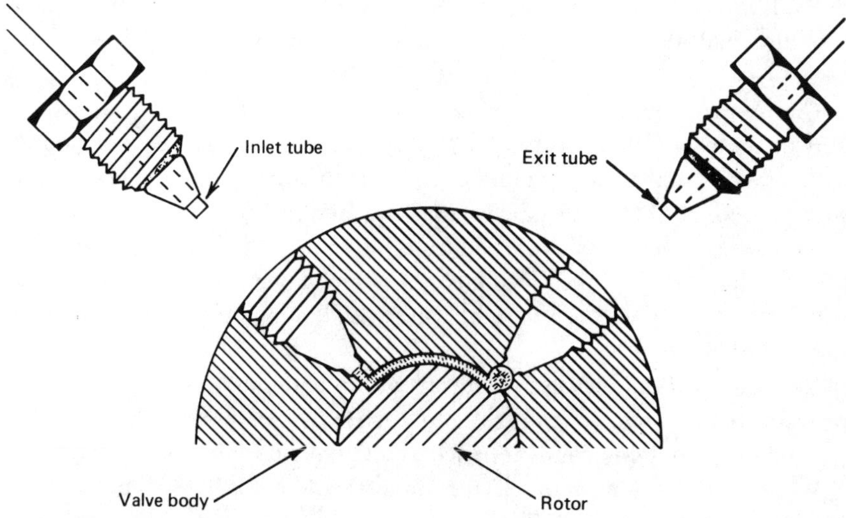

Figure 4.5 Zero dead volume fittings in an HPLC rotary sampling valve (courtesy of Valco Instruments Company).

Overtightening valve fittings, as well as any HPLC fittings, will produce several problems. First, a continuously leaking fitting can occur. Once overtightened, further torquing will not solve the problem. Second, loss of the nut, ferrule, tubing, and possibly the valve itself, which costs several hundred dollars, can also occur. Typically overtightening results in galled fittings where the threads are distorted. The threads in the valve body may be galled as well. Unless these can be rethreaded in your laboratory or shop, the valve body may be ruined or lost for repairs. Third, some fittings can be tightened so much, that the hexagonal fitting heads have been known to break off, leaving the male portion in the valve body. Unless this can be carefully drilled out, for example, with an "easy out" from a local hardware store, this valve body may be lost. Finally the user loses time. Downtime cannot be stressed enough as a loss to the chromatographer and his/her company or university. This is a constant source of frustration to users, yet so often it is brought upon by careless techniques.

4.4.4 Preparing Valve Tubing

The end of the small ID stainless steel tubing frequently used with valves will usually be prepared from one of two sources. One, when the tubing is first purchased, it will be a smooth, relatively flat end with the ID base of the tubing relatively consistent with the remaining bore throughout the tubing. Second, the end is prepared by the user who has cut his/her own piece. The condition and method of preparing the end of this tubing is important for several reasons.

The tubing should be butt centered and flat up against the inside of the port in the valve body for a good seal (see Figure 4.5). A piece of tubing without a perpendicularly flat end or with a bent and twisted end may never seal. Second, one of the major causes of scratched rotors and valve bodies are metal particles which are produced from the tube cutting operation and get between the rotor and the valve body. Any metal particles left behind from cutting the tubing can wash into the valve with the mobile phase and scratch the assembly.

New tubing should always be rinsed with mobile phase under operating pressure before being assembled onto any HPLC component. This can be washed into a suitable container such as a flask or beaker and discarded.

When cutting tubing yourself, it is important to cut it flat without closing the ID bore of the tubing. Tube cutting can be done with a file, hacksaw, lathe, or a commercial tubing cutter such as the knife file and rig from

Alltech Associates (catalog no. 7180) or the tube cut off machine from Scientific Systems, Inc., model TC-10, shown in Figure 4.6. The hacksaw tends to close the ID bore more than any other method. The most common technique is to score the tubing completely around the tube's outer surface, and then carefully bend the tubing with short back and forth motions about the score mark. This will weaken the tubing, and it will break apart similar to a glass rod. If a vise is used to hold the tubing, be sure not to mare the surface where a ferrule seal will be made. Once apart, file the end flat, being sure not to close the ID bore of the tubing at that end. Next use a no. 69 drill bit (0.029 inches) to ream out the ID bore, smooth the opening, and remove metal particles. Rinse the tubing out with solvent before using. These in-laboratory techniques take practice, and it is advised to practice on a piece of tubing to learn the process. Problems are more frequent with 0.009 inch ID tubing than with 0.020 or 0.040 inch ID tubing. Manufacturers traditionally use lathes or other on-line machine techniques to produce the fine ends on newly purchased tubing.

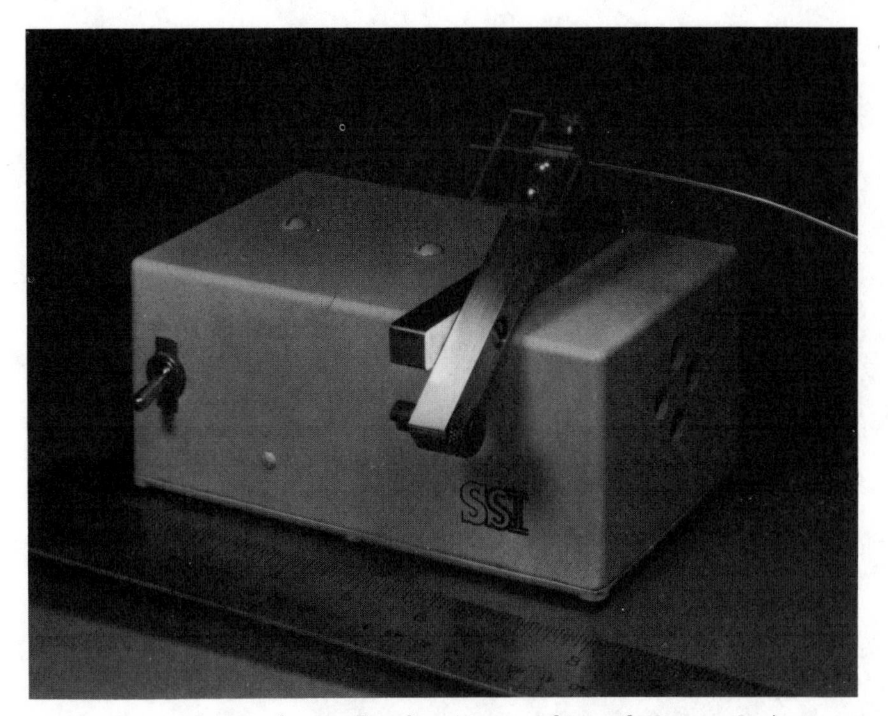

Figure 4.6 SSI tube cut-off machine (courtesy of Scientific Systems, Inc.).

4.4.5 Crossport Leaks

Most valves used are 6-port valves. More may be used depending on the system desired. Crossport leaks may start slowly and gradually increase in number and size. These leaks will ruin the precision of the injection and destroy any reproducibility and possibility for doing quantatitive work. Obviously a lack of reproducible peak heights with a valve may be due to crossport scratches.

Another way to identify these early is to use Teflon tubing for the exit line. A stainless steel fitting and ferrule can gently be used on Teflon, just as on stainless steel. The advantage of a Teflon line is its transparency. If a crossport leak occurs, the mobile phase can be observed passing through the line when the valve is in the load or run position. The drain is for sample excess during sample loop loading and manual sample loop rinsing. If liquid runs out from this line while running, there is a leak.

Lastly, the rotor can be dismantled and physically examined. The surface should be highly polished, smooth, and free from scratches.

4.4.6 Introducing Samples

Samples are introduced by syringes. However, these are not pressure syringes as with the syringe/septum injections described in Section 4.3. A typical Yale & Becten-Dickensen 1 ml glass luer tip tuberculin syringe can be used with an appropriate Kel-F male luer (Altex Scientific, Inc., or Glenco Scientific, Inc.) adapter, mounted into the sample entrance port of the valve.

For very small samples, Glenco Scientific, Inc. provides 10μl and 20μl syringes with a stainless steel injection port fitting for injecting accurate, reproducible samples into valve sample loops where very little sample is available. This allows, for example, a 10μl sample to be used to fill a 2μl loop.

It is important when injecting the sample into the loop to ensure that all the air is removed to prevent it from entering the system. This is done by flushing the loop with excess sample. When using a Teflon waste tube, air can be seen rinsing through to waste. It is also advised to expel all air from the syringe.

Never leave the syringe, particularly the luer tip type, in the valve overnight or for long periods of time. It is typical for the system sample port assembly to dry out and freeze up, making it difficult to be removed without breaking.

Always label the loop with its volume delivery, and never apply a side force perpendicular to the rotor shaft when turning the valve rotor. This can place a side force on the rotor and valve body, possibly distorting the polymeric rotor body and creating leaks. This can also occur by leaving the rotor only partially actuated.

4.5 OTHER INJECTION TECHNIQUES

Two additional injection methods need to be mentioned, the septum-less injector and automatic injectors.

Septum-less injectors, such as Waters Associates UK6, provide for pressureless syringe injection without septa and without interrupting the mobile phase flow. However, typical syringe care is required, and Teflon seals and valves can wear and leak. Consult the manufacturer for problems as a result of poor injection precision.

Automatic injectors are supplied by many of the HPLC manufacturers. These obviously offer the advantage of running many samples unattended. Most of these resemble the typical GC auto-samplers in concept. Proper positioning and sample vial labeling are critical. Depending on the sampler, sample vials have been known to be crushed or skipped out of sequence. For these problems consult the user's manual and the manufacturer.

4.6 SAMPLE CONTAINERS

HPLC allows very small samples to be injected. And just as evaporation can change a mobile phase ratio, it can also change the concentration of the sample while in use. Therefore it is very important when using a sample to keep it covered. This is particularly important when close tolerances on injection precision are required. Frequently 5 and 10 dram vials are used, and sloppy technique can ruin data and results. Vials with plastic retaining caps and rubber septa are also used.

Label the sample container with sample name, solvent, concentration, date and identification number to some notebook where additional information such as the sample's preparation may be found. Although a simple point, many a day's work has been lost by not adequately labeling samples and standards.

REFERENCES

1 E. L. Johnson and R. Stevenson, *Basic Liquid Chromatography*, Varian Associates, Palo Alto, CA, 1978.

2 D. Rodgers, *Developing an Analytical Method by High-Efficiency Liquid Chromatography*, Perkin-Elmer, Norwalk, CT, 1974.

3 K. J. Bombaugh in *Modern Practice of Liquid Chromatography*, J. J. Kirkland, Ed., Wiley-Interscience, John Wiley & Sons, Inc., New York, 1971.

4 L. R. Snyder and J. J. Kirkland, *Introduction to Modern Liquid Chromatography*, 2nd Edition, Wiley-Interscience, John Wiley & Sons, New York, 1979.

5 *Chromatographic Columns and Supplies*, Waters Associates, Inc., Milford, MA, p. 17, 1979.

6 *Septa and Septum Retainer for DuPont Liquid Chromatographs*, Product Bulletin 820PB3, DuPont Instruments, Wilmington, DE, 1971.

5
Columns

5.1 INTRODUCTION

The high performance column is probably the heart of the HPLC system. The development of this column technology in the late 1960s led to the evolution of the HPLC instrumentation and systems used today. These small particle columns (5–75 microns) have provided high efficiencies and a significantly higher number of theoretical plates, such as 10,000 plates/meter. However, the small particle diameters produced high back pressures, which prevented the use of the classical gravity feed approach and consequently resulted in the development of high pressure solvent delivery and sample introduction systems.

The earliest columns were the 30–75 micron solid core and thin porous layer particles. These pellicular packings had column efficiencies of 3000 plates/meter or more. Today's microparticulate columns with 5–10 micron particles have theoretical plate counts of up to 60,000 plates/meter (1). Yet with all this claim for efficiency, today's column technology, including packing techniques, care, use, and regeneration, is still in its infancy. This is evident by user problems such as lack of reproducibility from lot to lot of a single manufacturer's columns, the extremes in column lifetime from a few weeks to 6 months or more, and the lack of widely acceptable retention indices.

Although there is much to be learned about HPLC columns, there are many courses of action available to the user to prevent problems and extend column lifetimes. Many of the these are described below.

5.2 COLUMN SELECTION

Column selection is not always straightforward, and no established standard set of rules exist which will apply for every separation problem. Unless the chromatographer already has a ready library of separations (compound/column/mobile phase) available, various prescreening techniques such as TLC and trial and error methods will be needed.

Fortunately today's chromatographer is not working completely in the dark. Over the last 10 years many authors have discussed column selection techniques (1–9), and various manufacturers have published their own recommended easy to use guides.

The key to column selection when previous separation information is not available is knowing the chemistry of the sample. This is a critical area which so many chromatographers forget. Although the science of HPLC has developed to the point where one can often inject samples on randomly selected columns to find the right column to use, the proper place to start is with the simple principles of chemistry. The basics of wet chemistry and structure are too frequently treated lightly by today's student and user until hours of searching have been wasted and several columns contaminated or ruined.

Once the chemistry of the problem is taken into account, the following classical column selection steps can be used:

1 Determine the sample's molecular weight.
2 Determine the sample's range of solubility, both qualitatively and quantitatively.
3 Select a mode of separation based on the sample's molecular structure.

This is diagramed in Figure 5.1.

Knowing which column to select takes experience. Most column manufacturers supply descriptions of their columns in terms of solvent compatibility, classes of compounds separated, packing material characteristics, sample capacities, and so on. Excellent reviews have been published on HPLC columns, column packings, and their characteristics (10–12). Several characteristics of these packings are shown in table 5.1 (12).

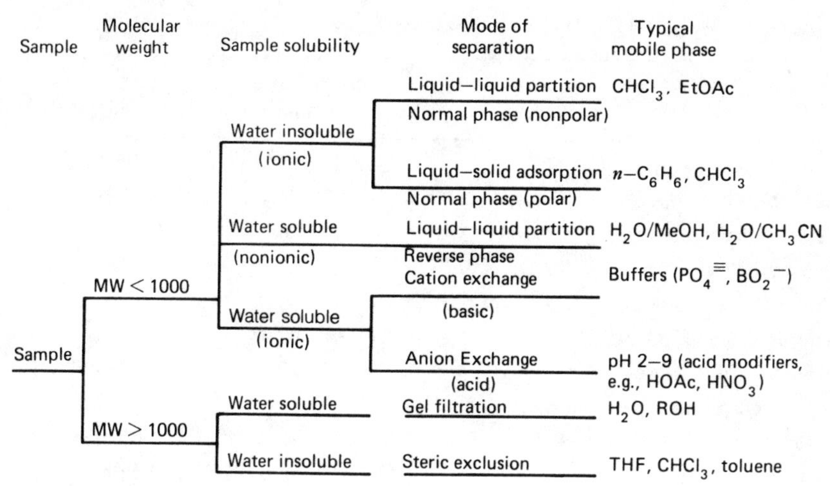

Figure 5.1 Typical column selection guide

5.3 COLUMN CARE

5.3.1 Solvents

As previously described in the chapter on mobile phase, knowledge and care in the choice and handling of the mobile phase solvents is critical toward an efficient HPLC system. A loss of column efficiency resulting in poor resolution, peak broadening, and loss of capacity can be attributed to solvent related conditions.

Column deactivation can occur with liquid-solid silica and alumina columns when solvents containing water and various preservatives or extenders are used. Ethyl acetate containing water, chloroform containing a low molecular weight alcohol, and THF containing butylated hydroxytoluene all can add to column deactivation (see Section 2.2.5). This will manifest itself in reduced resolution and selectivity.

Columns with nonpermanently bonded liquid phases can be deactivated with these solvents, destroying the liquid phase-solid support bonds. For example, water and low molecular weight alcohols will cleave the chemical bonds on such traditional liquid phases as OPN carbowax and C_8 on silica packings. These packings typically were advertised as nonbleeding packings. The early traditional columns using mechanical and weak adsorption forces to hold the liquid phase would often produce drifting baselines due to liquid

Table 5.1 Properties of HPLC Column Packings (12)

Property	Porous Layer Beads	Microparticles
Average particle size (μm)	30–40	5–10
Best HEPT[a] Values (mm)	0.2–0.4	0.01–0.03
Typical column lengths (cm)	50–100	10–30
Typical column diameters (mm)	2	2–5
Pressure drop (psi/cm)[b]	2	20
Sample capacity (mg/g)	0.05–0.01	1–5
Surface area (liquid-solid) (m^2/g)	10–15	400–600
Bonded phase coverage (%wt)	0.5–1.5	5–20
Ion-exchange capacity (μequiv/g)	10–40	2000–5000
Ease of packing	Easy, dry pack	Difficult, slurry pack
Cost		
Bulk packing	\$4–5/g (LS)[c] \$7–9/g (BP)[d]	\$2–3/g (LS) \$10–15/g (BP)
Prepacked columns	\$110–130 (LS) \$150–170 (BP)	\$200–250 (LS) \$225–275 (BP)

[a] HEPT—height equivalent to a theoretical plate
[b] Columns of equal dimension (2.1 mm ID) operated at 1 ml/min and mobile phase viscosity of 0.3 cp.
[c] LS—liquid solid.
[d] BP—bonded phase
(Reproduced from the *Journal of Chromatographic Science* by permission of Preston Publications, Inc.)

phase bleed and stripping. This was commonly overcome by adding a small percent of the liquid phase to the mobile phase or by using a precolumn with substantially higher loadings of the liquid phase. This would produce and maintain an equilibrium of the liquid phase to the solid support sites. Although this will work, extreme care is needed to adequately reproduce these system conditions from day to day and from laboratory to laboratory.

With steric exclusion columns improper solvent selection will affect the column efficiency by changing the column bed volume as a result of its solvent swell characteristics. This occurs with styrene columns. Using glass beads, such as CPG columns (ElectroNucleonics, Inc.), will avoid this problem. However, be aware of competing adsorption processes that can occur with these silica based particles. Deactivation can be accomplished by increasing the pH and the ionic strength of the mobile phase (13), adding 1% ethanol to the mobile phase (13), silanizing the glass beads with dimethyldichlorosilane (DMDCS) (14), or treating the glass beads with polyethylene oxide (15). Also be aware of microbiological growth in these columns when using aqueous systems. These systems are very susceptible to microbiological colonies which will clog the packing pores, increasing the back pressure and destroying its separation power.

The introduction of the permanently bonded liquid partition phases eliminated for the most part the problem of liquid phase bleeding. However, the pH of the solvent media is as critical with these packings as with the earlier chemically and mechanically bonded packings. The usual mobile phase operating pH range is 2.0–8.0. At pH greater than 7.0, silica begins to exhibit solubility. Silica is soluble at pH 7.5 in the ppm range in CH_3CN/H_2O and $MeOH/H_2O$ systems, and it reaches part per thousand solubility at pH 8.5. As silica dissolution increases, the column efficiency drops. Voids will develop in the column, producing band broadening, loss of resolution, and split peaks (see Section 5.4). On the extreme acid side, pH below 2.0, the silicon-carbon bonds between the solid support and the liquid phase will cleave. This has the same effect as water and low molecular weight alcohols have on the earlier columns with functional group bonding such as with ester linkages.

5.3.2 Filters

A major problem with column deterioration is particulate and chemical contamination buildup at the head of the column. The buildup results in increased back pressures from physical flow restriction or peak distortions due to extra column adsorption effects (see Section 5.4.2).

This can be prevented or reduced by properly filtering solvents and samples before use and by carefully choosing solvents in which the sample will be soluble. However, even with these precautions, contamination can still build up at the head of the column. Particulates can still pass through the filters, and chemical adsorption of impurities or unknowns in the sample can still occur.

Most columns today have inlet and outlet filters. These are primarily present to prevent the column packing material from escaping from the column. The frit pores are usually smaller than the mean particle diameter of the packing material. In some of the earlier DuPont columns the inlet filter had frit pores larger than the outlet. The inlet filter served only as a holding device for the packing and as a coarse particle filter. Reversing the flow direction through the column dispelled the packing material into the detector.

Column end frits are made of stainless steel. They are either embedded in the column end fitting by pressure or spring loading or they are pressure fitted into the column end fitting itself. Some columns have been plugged with glass wool. This works satisfactorily for larger particles between 30 and 75 microns and for operating pressures under 1500 psi. Glass wool is also easy to handle with columns that are dry packed. However, the glass wool can provide adsorptive sites for samples. Even silanized glass wool must be used with caution. Another adsorptive material that has been used with commercial stainless steel frits is a small circular piece of filter paper placed between the column and the stainless steel frit (Waters Associates). The idea is to use the filter paper to retard contaminants or particulates on the filter disk rather than on the column packing itself. Once the paper disk becomes plugged or contaminated, it is easier to simply replace the disk than the column packing material. These disks can become easily contaminated and will retard sample and mobile phase flow. The user must be aware of the presence of such filter disks and replace them frequently. The idea of a disposable disk is good, but it can be more trouble than it is worth. The problems described in Section 5.4.2 will still occur with these filters. It is often recommended to discard these disks and use the columns without them.

The user must determine what type of frits and end fittings are present on each of his/her columns in order to clean or replace them. When the back pressure on the column increases, it is common to remove the end fitting and frit and clean them in a suitable solvent such as 6*N* HNO_3 under sonication. Those that cannot be removed can be cleaned by back flushing the column. However, this is recommended only as a last measure. Back flushing can easily distort or disrupt the column bed, causing voids or changing the packing geometry. This will reduce the number of theoretical plates and can change the column performance characteristics so that it will not perform satisfactorily or similarly again.

Before removing any column end fittings be sure to check with the manufacturer's literature. There are often special instructions including notification of voiding any column warranties. Brownlee Labs has even epoxied its end fitting onto the column and notified the user that breaking the epoxy seal voids any guarantees by them for that column.

5.3.3 Daily Column Care and Preservation

The chromatographer with CPMA develops certain routine procedures which further column lifetime and can aid in troubleshooting the HPLC system. This section lists daily do's and don't's for proper column care.

1 During startup each day, gradually increase the flow rate and column pressure. Slowly increasing the pressure will avoid pressure shock to the column bed, thus preventing distortion and voids.
2 Allow the HPLC system to come to equilibrium before injecting samples. This can be determined by watching the detector recorder baseline. It allows the column to become equilibrated with the mobile phase.
3 Before injecting samples, always feel the HPLC fittings with your hand, including the column end fittings. This is a good indication of a leak. Leaks are not always evident by mobile phase dripping from them. Many times a leak is so small that it can only be detected by coolness to the touch due to evaporation at the fitting. Even such a small leak can cause baseline drift, introduction of air, and it can interfere with peak height and area reproducibility for quantitation.
4 Never overtighten the column end fittings at either the column nut or the connection to the solvent lines as mentioned earlier (see Section 4.4.3). Too much torque can gall the fitting threads and cause a leak that further tightening will not stop. If a leak is observed, whether by actual solvent observed or by coolness to the touch, a slight tightening of the fitting can be applied. If this does not stop the leak, back off on the fitting and retighten. When doing this, be sure that the fitting and the column or tubing are securely and squarely butted up against the fitting and ferrule first. Never use wrenches with a long lever arm. Short wrenches are preferred so that too much torque is not mistakenly applied. A small torque wrench would be preferred.
5 Do not place the column in drafts or in direct sunlight. Drafts and sunlight will cause temperature gradients across the column, which will appear as drifts on the detector-recorder printout. If drifting does occur and temperature gradient effects are suspected, protect the column with some form of thermal insulation. This can be done by placing the column in a portable column oven compartment, wrap-

ping the column and its end fittings with a cloth or towel, or covering the column and its end fittings with rubber tubing or hose. The latter is done by splitting a proper length of tubing or hose lengthwise and fitting it over the column lengthwise. Instruments with enclosed column compartments (such as Hewlett-Packard, DuPont, Perkin-Elmer, Spectra-Physics, Varian) usually do not suffer from draft related problems. Whether or not drafts are an immediate problem, keeping the column thermally protected is good CPMA.

6 When carrying out quality control analyses, use columns dedicated to a single analysis only. Although this requires a separate column for each type of routine analysis, in the long run it will prevent column contamination and mixups, and it will probably extend the lifetime of the column since it is used with only one system.

7 When injecting samples that add contaminants to the column, it is recommended to flush the column with a suitable solvent overnight. This can be done with a few hundred milliliters at a slow flow rate so as not to use extremely large amounts of expensive solvent. In the morning reequilibrate the column with the mobile phase. This may take up to 30 min. If the instrument is programmable, program it to switch solvent reservoirs in the morning. Put the flush solvent in reservoir A and the mobile phase in reservoir B.

8 When removing, replacing, or storing columns, handle them gently. Do not bang, bump, or physically mishandle them. This abuse will physically shock the column bed, distorting it and creating voids and channels. Like the end fittings, the column must be treated gently. Columns should be handled like fine china.

9 Store columns labeled as discussed in Section 5.3.4. Keep used columns separated from new columns. Do not store columns under temperature extremes as the contraction and expansion of the column can create channels. Always store the column wet. Column beds left to dry out tend to shrink, crack, and develop voids and channels.

10 Always store an adequate supply of back up columns.

5.3.4. Column Storage and Labeling

Columns should always be stored wet and never allowed to go dry. A dry column will eventually have voids because the packing will shrink away from the column walls, or because of shrinking within the packing itself. These voids produce column bed distortion which can result in band broadening, loss of resolution, and split peaks (see Section 5.4). When the column is no longer needed, disconnect it from the HPLC system, and cap it securely.

Never store the column with buffer solutions or salts, particularly halogen salts, since they will corrode the stainless steel. If potentially deactivating solvents are used or solutes from the sample are known to build up on the column, flush the column with a suitable solvent before storage. Use dry solvents for storing liquid-solid adsorption columns. Water, water-MeOH, or water-CH_3CN are all used for reverse phase columns. Although none are discouraged for storage, the use of 100% organic solvent for storage in these columns is preferred to reduce the slightest dissolution of the silica support. Store ion-exchange and steric exclusion columns in solvents compatibile with their ionic behavior and swell characteristics respectively.

Except for steric exclusion columns, there is little consensus over storing columns under pressure. This is done to keep the column wet and prevent shrinkage. Most users do not follow this procedure, and it is one that the chromatographer will have to determine for himself/herself.

Columns should not be stored in damp areas or under extremes in temperature. There are differences in the coefficients of expansion between the stainless steel column and the packing, and repeated expansion and contraction of the column can distort the column bed. Maintaining the temperature between 15 and 30°C all year long can preserve the column lifetimes well over a year.

A major problem in HPLC laboratories is the lack of adequate column labeling and history. It is not uncommon to open a laboratory bench drawer and find unlabeled columns from many manufacturers produced over many years. Their use becomes either guess work relying on memory or a loss because they have to be discarded.

Every column must be labeled. This can be done with commercially available aluminum tags as used in GC. Label tape can also be used. The tags should include:

Liquid phase
Solid support
Manufacturer
Last date used
Last solvent used

The last item may or may not be the storage solvent. In this case the storage solvent should also be noted. This information will assist the next user in the proper choice of column and the correct startup conditions.

Some columns have tags identifying the column inlet and/or outlet. These should not be removed. If no tag or marking is available, mark the inlet or the direction of flow. This will prevent accidentally back flushing a column or flushing dirt off the head of the column into the detector.

It is good CPMA practice to keep a column log. This becomes more important as the number of columns and users increases in any given laboratory. Although it takes a few minutes each day to maintain a log, it provides a column history that can save hours of redevelopment work and troubleshooting. Several types of logging procedures can be used. A single page per column can be used (Figure 5.2) where there are many columns, or a multiple column log sheet can be used (Figure 5.3) where only a few columns will be listed. Most columns come with a serial number attached or scribed onto them. If there is none, add your own by marking the column end fitting or labeling the column. This becomes important when many columns of the same type are used. It is also advisable to keep a list of all the columns purchased and to note both the date received and the date discarded. Columns whose lifetimes have ended should be discarded or returned to the manufacturer clearly marked. They should not be stored

Manufacturer ______________________ Serial Number ______________

Date Received ______________________ Length __________

ID __________ OD__________

Date Used	User Identification	Sample	Mobile Phase	Left in Instrument Overnight	Storage Conditions	
					Solvent	Date

Figure 5.2 Individual column log.

Date	User Identification	Column Description	Column Serial Number	Sample	Mobile Phase	Storage Condition Solvent	Storage Condition Date

Figure 5.3 Multiple column log.

with the good columns. Knowing column lifetimes for your laboratory's needs will assist in analysis and inventory planning.

5.4 COLUMN DETERIORATION AND PEAK DISTORTION

Loss of resolution and peak distortion are problems that the chromatographer encounters frequently. These problems are often column related, although this is not always readily evident. Frequently these problems are caused by column bed distortion and/or contamination.

5.4.1 Column Bed Distortion

A column bed that has cracked, dried out (even partially), or shrunk will cause resolution problems. This occurs when the sample is prevented from starting on the column as a tight slug. Instead it will enter the column, spread out and be diluted.

Column bed recession can be caused by sudden pressure shocks to the system, for example, turning the pumping system suddenly on and off at full operating pressure. It is preferred that the system flow rate be built up or reduced gradually when the system is turned on and off. One may not always have problems when using sudden pressure drop changes, but again it is good CPMA. Recession can be caused by physical shocks from banging or bumping the column or can also develop gradually with normal use. This latter type of bed settling occurs because of the particle size distribution and the packing array. As the bed is physically distorted under normal operating pressures or chemical attack, packing particles can shift or redistribute themselves into a smaller packing pattern. As this occurs, the bed can shrink. Since it is under pressure at the head of the column, it will tend to be compressed toward the column outlet, producing a void at the head of the column.

Chemical deterioration can also cause recession. For example, if the packing slowly dissolves in the mobile phase, as silica will do at high pH, the gap left by the particles becoming smaller can be filled by other particles. This can result in a continuously cascading effect until a void at the head of the column is produced. The user can avoid or minimize this by recognizing the chemistry of the systems being used. Some dissolution can always occur, but proper care can minimize this.

Poorly prepared and/or packed columns can also do this. If the packing technique is not complete or the particle distribution is too wide, it is easier for segregation to occur.

The effect of column bed recession usually is gradual, and it is observed with the detector printout by peak broadening, a loss in resolution, and/or peak distortion. If this is suspected, the user can confirm it by disconnecting the column and carefully removing the column end fitting at the head of the column and looking at the column bed. If the column bed has receded, one of several courses of action can be taken:

1 Discard the column.
2 Repack the entire column.
3 Repack the void portion of the column.

The action taken will depend on the user's column supply, ability to pack HPLC columns, supply of extra packing materials, time, and money. Repacking the void is discussed further in Section 5.5. In any event, the receded column should not be used as is.

Column bed distortion manifested by a void at the head of the column is the easiest to detect and correct. Voids and cracks in the interior of the column bed cannot be seen with today's HPLC column. The classical gravity fed glass columns allowed the user to see visually a major portion of the column bed, and to determine whether this is occurring. With today's column technology transparent columns are no longer used, and the only method of detecting these problems is through the detector printout or by visually examining the column bed ends. A column bed that has dried out and cracked will produce channels particularly by shrinking and pulling away from the column walls. This will distort the solute bands, causing band broadening and loss of resolution. It can also split the solute band causing the peaks to split giving the appearance that another compound is present (see Section 5.4.2). If damage in the column interior is suspected, another good column should be tried to determine whether column deterioration is the problem.

5.4.2 Effects of Column Impurities

Impurity buildup on an HPLC column can have several negative effects, ranging from increasing the column back pressure to producing spurious and split peaks. This form of contamination results from improperly prepared samples, sample and solvent contamination, and from introducing the wrong sample or solvent combinations onto the column.

It is not uncommon to dissolve the sample in the mobile phase or a different miscible solvent and have some of the sample components precipitate out onto the head of the column (see Section 2.2.4). This can be the solute of interest or some other compound which becomes adsorbed onto the head of the column. Particulate or lint contaminants can also be deposited onto the column if sample and solvent filtering is inadequate. In these cases the contaminants are not soluble in the sample solvent or the mobile phase.

Such contamination reduces resolution and produces peak distortion. Loss of resolution is principally observed with peaks where α is small and one of the peaks is considerably smaller than the other. This is illustrated in Figure 5.4.

A more difficult problem to recognize is the development of additional peaks. With the proper set of conditions, column contamination or bed distortion can split a solute band, producing two peaks of the same compound. This is particularly misleading during method development where unknown peaks may be expected. It can manifest itself by producing an extra peak on one or all of the normally eluting peaks. It typically will appear as a well defined shoulder (Figure 5.5). Since the shoulder is so well defined, the user may not consider the second peak to be the same solute material as the first. The author has experienced this phenomenon. These peaks can be identified by column collection followed by mass spectroscopy.

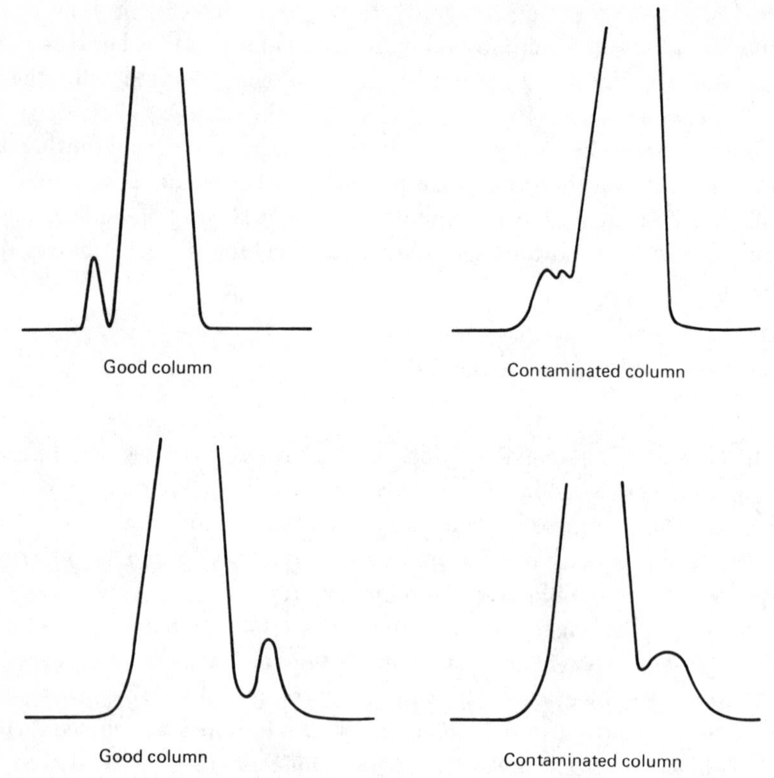

Figure 5.4 Peak distortion from on-column impurities.

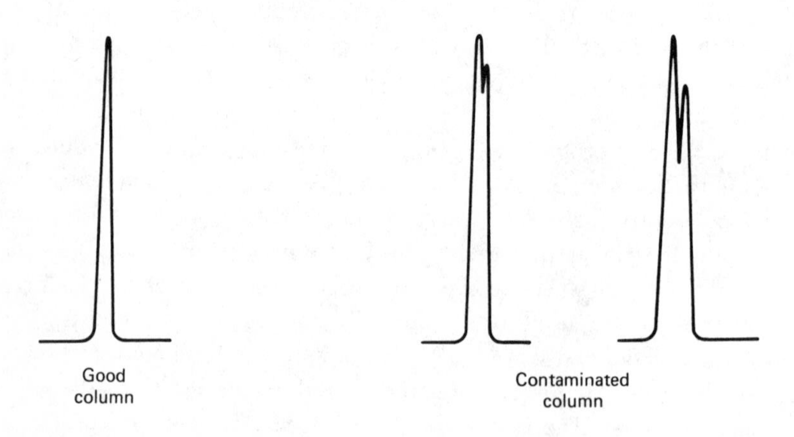

Figure 5.5 Spurious peaks from on-column impurities or column bed distortion.

5.5 COLUMN REGENERATION

The emphasis with column handling is on prevention, as dictated by good CPMA. However, as discussed earlier in this chapter, many problems can develop that will damage the column and affect column performance, such as impurity buildup, chemical attack, or column bed cracking. The ability to regenerate a column where performance has dropped can greatly save time and money. Several restorative techniques are described below.

5.5.1 Column Contamination

A good measure of column performance is to run the test mixture recommended by the column manufacturer or a mixture oriented toward the user's typical separations. Loss in efficiency is often the result of irreversible adsorption of contaminants on the column bed. These adsorbed species will tie up active column sites, reducing the column's separation power.

The simplest method of regenerating a normal phase column contaminated in this manner is to pass 40–60 column volumes of appropriately stronger eluting solvents through the column. For normal phase columns it is preferred to start with the least polar solvent first and try the test mixture between selected washings. This takes time at first, but it can prevent long reequilibration times with strongly polar mobile phases. A solvent as strong as methanol can be used. However, bringing the column back to equilibrium with the mobile phase such as $CHCl_3$-C_6H_6 will take time since MeOH can deactivate the column. It is done by passing backwards through the elutropic series. Be sure to use dry solvents since H_2O will tend to substantially deactivate the column.

Reverse phase columns can be washed similarly with solvents like MeOH, THF, or CH_2Cl_2. The same approach as with normal phase columns can be used, except that the opposite polarity rules apply. In either case it is necessary to pass the appropriate solvents from the elutropic series through the column to return to the original condition.

To remove biological materials such as fats and oils from reverse phase columns wash with CH_3CN. During elution inject several 1 ml volumes of DMSO. Caution is necessary since the higher viscosity of DMSO and a high flow rate can cause bed compression. It is advisable to reduce the flow rate during these injection periods (16).

A simple set of regeneration guidelines for a variety of columns is presented in Figure 5.6.

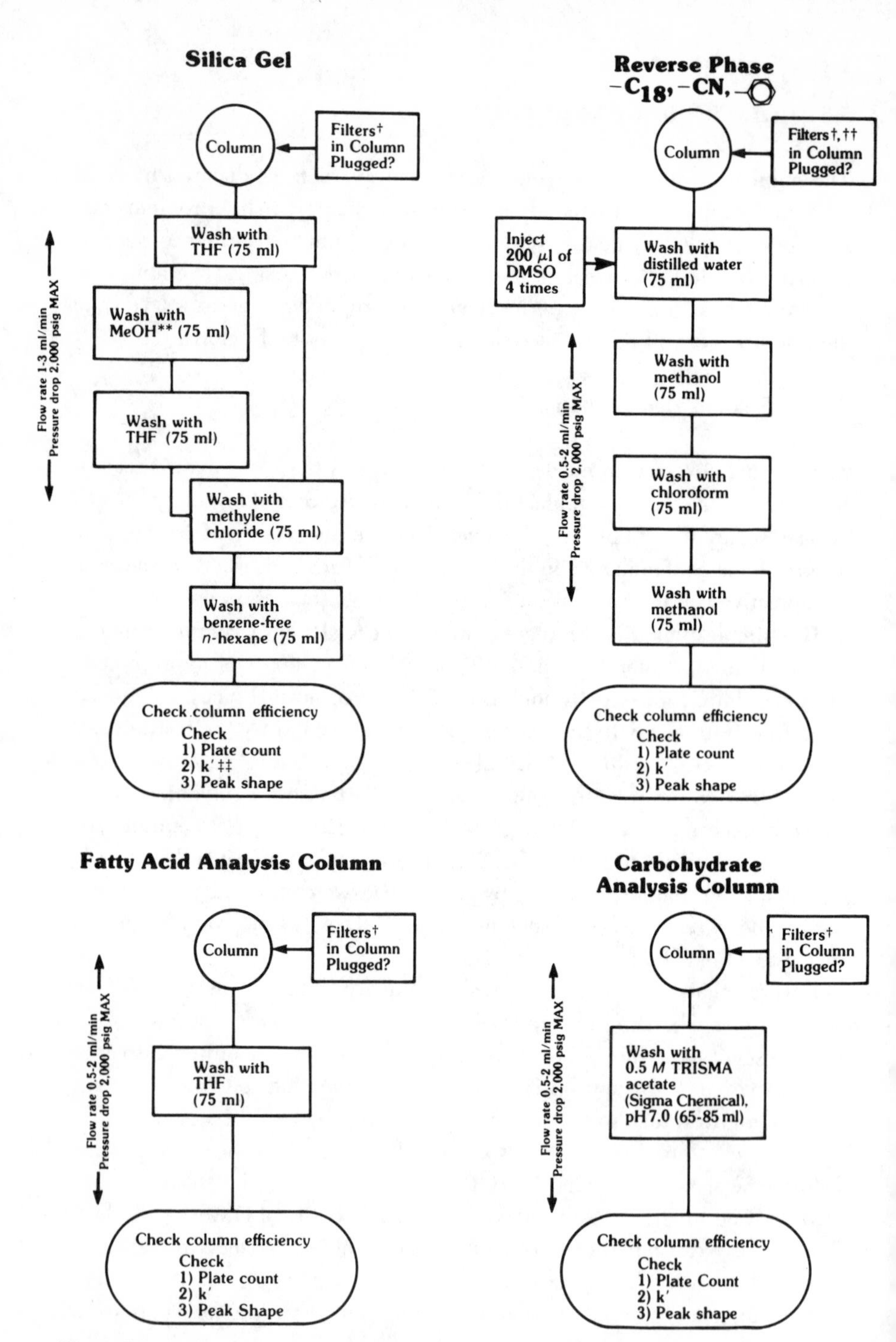

Figure 5.6 Regeneration of packed columns (courtesy of Waters Associates, Inc.).

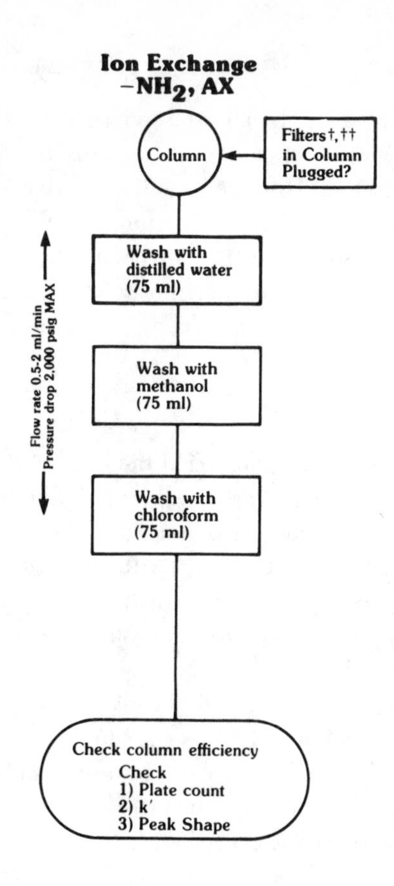

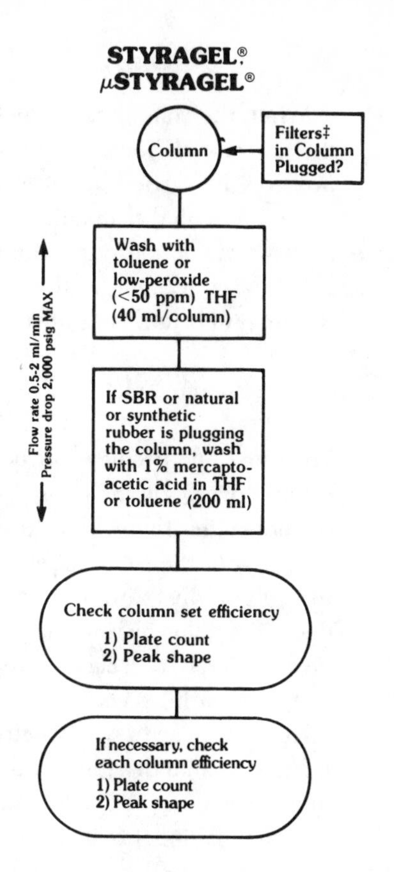

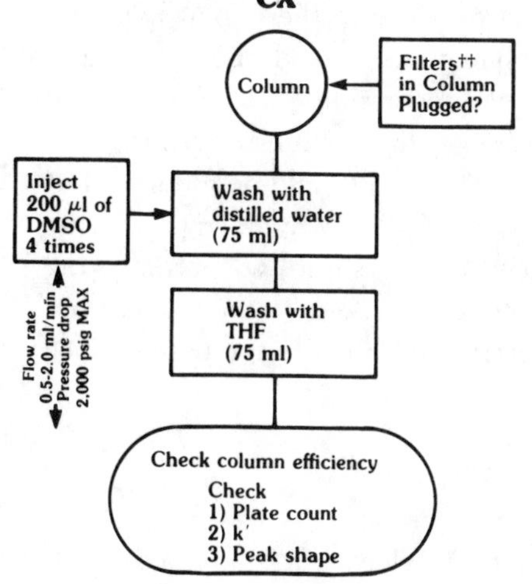

Footnotes

***If a column becomes contaminated (material is not eluted), regeneration may return the column to a useful tool. However, there is no guarantee that regeneration will return the selectivity and pressure drop of a packed column to their original values.**

The solvents given in this chart will work in most cases. However, if you know what material is contaminating the column, pick a solvent (compatible with the column) that has good solubility for the contaminant. Flushing with this solvent should elute the retained material.

†There is a replaceable 2-μ filter in the hex nut of the 4 mm ID columns. P N-84095. Replacement is recommended. To clean, place the filter in a sonic bath with a small amount of detergent.

††On the 2-mm ID columns BONDAPAK AX and CX, the filter is part of the end fitting. Clean by placing the end fitting in a sonic bath with a small amount of detergent or replace P N-27249.

‡The end fittings cannot be removed from μ STYRAGEL columns. You can back flush if necessary.

‡‡k' will vary with the water content of the *n*-hexane

Methanol may be required if a component is very strongly adsorbed on the silica surface. There are some indications that mixing 2,2'-dimethoxypropane with the methylene chloride wash will assist in removing small amounts of water from silica columns.

D67

When the column is washed, the effluent can be allowed to pass through the detector to permit monitoring the elution of contaminates. This would probably be carried out at lower sensitivities. However, caution is advised since any eluted contaminates could contaminate the detector cell. While returning to equilibrium with either increasing or decreasing polar solvents, it is advised to monitor the column effluent with the detector to determine when equilibrium is achieved. This is evident by a nondrifting baseline.

5.5.2 Column Inlet Voids

The cause and problems associated with voids developing at the head of the column were discussed in Section 5.4.1. They are easily observed by disconnecting the column from the HPLC system and removing the inlet fitting, with the column held securely in a vertical position with the head of the column in the top position. If the void is no greater than about 1 cm deep, the following procedure can be applied. Otherwise the column should be completely repacked or replaced.

A column inlet void can be repacked with either glass beads or packing material similar to what is already in the column. Particles greater than 20 microns can be dry filled with the tap fill method. For particles less than 20 microns a slurry is made with a column compatible solvent. The slurry is added in increments. Each increment is allowed to settle before the next one is added. Once the void is filled, level off the packing with a spatula, replace the column inlet fitting, and reconnect the column to the HPLC system. Slowly increase the flow rate and system pressure to the upper operating limit. Pass 10–20 column volumes of mobile phase through the column and then gradually reduce the flow rate and pressure to zero. Remove the column from the HPLC system and remove the column inlet fitting as above. Visually examine the column inlet. The void should be reduced, but it may not be totally filled due to compression of the newly added packing by the system's operating pressure. Repeat this process until no more packing can be added to the column. Once the column is packed, it is advised to run a test mixture through the column to determine if the column performance has been satisfactorily improved.

5.5.3 Column Inlet Contamination

It is common for impurities or contaminants to collect at the end of a column. The effects of such a buildup were discussed in Section 5.4.2. The presence of contamination is frequently detected by a visual inspection after having removed from the HPLC system the column and the column inlet fitting as described in Section 5.5.2. The column bed will appear discolored, being brown, yellow, or darker in appearance.

In this case carefully remove the upper layer of packing with a spatula until all the discolored packing is gone. Then repack the column as described in Section 5.5.2. If the discolored packing exceeds 1 cm in depth, the column should be completely repacked or replaced.

5.5.4 Thermal Restorative Methods

Two thermal regenerative procedures have been described in the literature (16). Both are superior procedures, but require more equipment and time to carry out. The procedure described is for normal phase columns.

Method 1 (Using a Programmable GC)

1 Pass organic solvents of gradually decreasing polarity and increasing volatility through the column. Be sure each solvent has come to equilibrium.

2 Remove the column from the HPLC unit and slowly pass an inert gas (He, N_2, Ar) through the column at 5–10 ml/min. This removes all traces of volatile solvents. *Caution:* Do this in a laboratory hood away from flames and hot plates.

3 Place the column in a programmable GC oven. Attach N_2 gas at a slow flow rate to the column inlet and increase the oven temperature by 2–3°C/min up to 150°C.

4 Hold at 150°C overnight with the N_2 carrier on.

5 Turn the oven off the next morning and allow the column to cool gradually to room temperature in the GC oven with the oven lid closed. Removing the column while warm or hot can cause voids and channels to develop due to differences in the coefficients of expansion between column walls and packing material.

6 Reconnect the column to the HPLC system and pass a nonpolar solvent like C_6H_{14} or C_8H_{18} through the column for 1–2 hr at 1–2 ml/min.

7 Slowly pass more polar solvents through the column, allowing each to come to equilibrium with the column. This can be monitored by the detector.

8 Stop with the final solvent desired. Pass this through the column for about 1 hr.
9 Check the column with an appropriate test mixture for resolution and number of theoretical plates.

Method 2 (Using an HPLC Oven)

1 Pass organic solvents of gradually decreasing polarity and increasing volatility through the column. Be sure each solvent has come to equilibrium.
2 Stop the solvent flow.
3 Disconnect the column outlet.
4 Increase the oven temperature slowly to 150°C over several hours.
5 Hold at 150°C overnight.
6 Turn the oven off the next morning and allow the column to cool gradually to room temperature in the HPLC oven with the oven door closed. Removing the column while warm or hot can cause voids and channels to develop due to differences in the coefficients of expansion between column walls and packing material.
7 Reconnect the column to the HPLC system.
8 Pass hexane through the column.
9 Slowly pass more polar solvents through the column, allowing each to come to equilibrium with the column. This can be monitored by the detector.
10 Stop with the final solvent desired. Pass this through the column for about 1 hr.
11 Check the column with an appropriate test mixture for resolution and number of theoretical plates.

5.5.5 Column Repacking

Columns that are badly deteriorated must be completely repacked or replaced. Columns with packing materials larger than 20 microns can be dry packed with the tap-fill method. For column materials smaller than 20 microns the slurry pack method must be used. Slurry packing equipment can be purchased from several suppliers (Table 5.2), and the technique has been described in the literature (17–18).

Table 5.2 Suppliers of Slurry Packing Devices

Alltech Associates 202 Campus Drive Arlington Heights, IL 60004
Micromeritics Instrument Corporation 5680 Goshen Springs Road Norcross, GA 30093
Rainin Instrument Company, Inc. Mack Road Voburn, MA 01801

Packing columns can save costs. However, to make it worthwhile, the technique needs to be learned and a solid quality control program for testing the columns must be developed and instituted. This is not a technique to be used once in a while. In this case purchasing new columns is recommended since occasionally packing columns will probably lead to more lost time in trying to perfect the technique each time it is needed. The optimum approach would be for a laboratory to establish a column packing group who packs, repacks, and checks out the columns for the entire laboratory. This can be preferred to purchasing columns not only because the column hardware can be reused, but these columns can be controlled to specifications tailored to the user's needs.

REFERENCES

1 J. J. DeStefano, A. P. Goldberg, J. P. Larmann, and N. A. Parris, *Industrial Research & Development,* 99–103 (April, 1980).

2 L. R. Snyder and J. J. Kirkland, in *Modern Practice of Liquid Chromatography,* J. J. Kirkland, Ed., Wiley-Interscience, John Wiley & Sons, Inc., New York, 1971, Chapters 6 and 8.

3 L. R. Synder and J. J. Kirkland, *Introduction to Modern Liquid Chromatography,* 2nd Edition, Wiley-Interscience, John Wiley & Sons, Inc., New York, 1979.

4 E. L. Johnson and R. Stevenson, *Basic Liquid Chromatography,* Varian Associates, Palo Alto, CA, 1978.

5 Column Selection in Liquid Chromatography, *Spectra Physics Chromatography Review,* 3(1) (1977).

6 G. J. Fallick and J. L. Waters, *American Laboratory,* **4**(8) 21 (1972).

7 L. R. Snyder, *Journal of Chromatographic Science,* **10** (4) 200–212 (1972).

8 L. R. Snyder, *Journal of Chromatographic Science,* **10** (6) 369–379 (1972).

9 L. R. Snyder, *Journal of Chromatographic Science,* **15**(10) 441–449 (1977).

10 R. Majors, *American Laboratory*, **4**(4) 27–39 (1972).

11 R. J. Laub, *Research and Development*, **7**, 24–28 (1974).

12 R. Majors, *Journal of Chromatographic Science*, **15**(9) 333–351 (1977).

13 *Controlled Pore Glass*, ElectroNucleonics, Inc., March 1972.

14 Personal communication from ElectroNucleonics, Inc.

15 C. W. Hiatt, A. Shelokov, E. J. Rosenthal and J. M. Galimore, *Journal of Chromatography*, **56**, 362–364 (1971).

16 I. S. Krull, U. Goff and R. B. Ashworth, *American Laboratory*, **10**(8) 31–36 (1978).

17 E. J. Kikta, Jr., A Portable Slurry Packing Apparatus for High Performance Liquid Chromatography, *Journal of Liquid Chromatography*, **2**(1), 129–144 (1979).

18 R. McIlwrick, More about How to Pack Your Own LC Columns, *Spectra Physics Chromatography Review*, **3**(1) 5 (1977).

6 Detection Systems

6.1 INTRODUCTION

The detector is probably the most important component of the HPLC system with regard to maintenance and troubleshooting. Besides its obvious usefulness for detecting solutes for both qualitative and quantitative measurements, the HPLC detector serves as a troubleshooting device itself. Its ability, for example, to respond to flow variations, gas bubbles in the pumping system, leaks, temperature gradients across the instrument, and fluctuations in back pressure make the detector a valuable measuring device of the system's operating performance. A detector's sensitivity to these changes is usually considered a liability to the average user. However, the chromatographer who is aware of these effects and what causes them can put these signals to valuable use in maintenance and in correcting system problems.

There are probably more types of detectors than there are variations for any other component of the HPLC system. Nearly every form of qualitative detection and quantitative measurement available to the analytical chemist has been tried and/or adapted to HPLC. This ranges from the classical methods of effluent collection followed by gravimetric, volumetric, or other instrumental methods of measurement to simply adapting a particular analytical instrument to be an integral component of the HPLC system. This has opened the door for the user to measure numerous properties of solutes of interest, and it makes method development and analysis easier. Table 6.1

lists many of the types of detectors used for HPLC, both those commercially available and those developed for specific research interests. Scott (22) Karasek (23) and Wise and May (24) have reviewed many other new and experimental detectors.

Detectors are typically classified as either (1) bulk property or universal detectors or (2) solute property selective or specific detectors. The bulk property detectors have a wide range of applications, particularly for classes of compounds where a specific detector is not available or applicable. These detectors measure the solute as a function of change in some property of the mobile phase. The solute property detector has a narrower range of application. These detectors measure a specific property of the solute itself

Table 6.1 Characteristics of Some HPLC Detectors

Detector	Classification	Sensitivity (g)	Reference
Refractometer	Universal	10^{-6}	1,2,3
Ultraviolet-visible	Selective	10^{-9}	3,4,5
Fluorescence	Selective	10^{-10}	3,6,7,8
Electrochemical	Selective	10^{-12}	9,10
Infrared	Selective	10^{-6}	1,11
Solute transport[a]	Selective		
Moving wire-FID		10^{-6}	12
Moving wire-EC		10^{-10}	13
TLC		10^{-10}	14
Mass spectrometry	Universal	10^{-9}	15
	Selective (selected ion mode)	10^{-12}	15
Atomic absorption	Selective	10^{-9}	16
Flame photometry	Selective	10^{-7}	17
Thermal conductivity	Universal	10^{-8}	1,6
Permittivity (Differential Capacitance)	Universal	10^{-8}	18
Evaporative analyzer	Universal	10^{-3}	19
Inductively coupled plasma	Selective	10^{-9}	20
Nitrogen detector	Selective	10^{-9}	21

[a]FID—flame ionization detector, EC—electron capture

which is not characteristic of the mobile phase. By virtue of their specificity, the selective detectors are more sensitive and usually less affected by variations in the mobile phase. Detector classification is sometimes broken into destructive and nondestructive techniques as well.

For maintenance and troubleshooting only the most popular detectors will be discussed in depth, namely, the refractive index and the UV-visible detectors. Although many detectors exist, nearly every user of HPLC has worked with either the refractive index and/or the UV-visible detector. The principles surrounding the use of these detectors for purposes of troubleshooting HPLC system problems as well as the care and maintenance of these specific detectors can be extrapolated in many cases to other HPLC detector systems. A summary of the specifications for the detectors to be discussed here has been compiled by Snyder and Kirkland and is given in Table 6.2 (1).

When selecting a detector for a particular study, the user should keep in mind many characteristics besides sensitivity. Similarly to selecting the proper column (see Section 5.2), proper understanding and selection of a detector should be part of the user's CPMA, which can save the chromatographer hours of misdirected time. Table 6.3 lists many characteristics the user should consider before choosing a detector.

6.2 TERMINOLOGY

There are some terms used in reference to detector characteristics that are common to nearly every HPLC detector. Several of these terms are briefly described below and are referred to throughout this discussion and in the appendices. Although many of these terms are usually considered in a negative connotation, their mere presence can help the user identify problems.

6.2.1 Noise

Noise is basically the variations in the detector signal observed with a strip chart recorder when no sample is passing through the detector cell. Ideally a detector with only mobile phase passing through it and no temperature, pressure or flow fluctuations should draw a straight line parallel with the detector zero line. However, the user will seldom experience the ideal, and since noise is either random or continuous, it is not always easily identified.

Table 6.2 **Typical Specifications for most-used LC Detectors (1) (Reprinted with permission of the publisher.)**

Parameter (units)	UV (Absorbance)	RI (RI units)	Radioactivity	Electrochemical (μ amp)	Infrared (absorbance)	Fluorometer	Conductivity (μMho)
Type	Selective	General	Selective	Selective	Selective	Selective	Selective
Useful with gradients	Yes	No	Yes	No	Yes	Yes	No
Upper limit of linear dynamic range	2-3	10^{-3}	N.A.[a]	2×10^{-5}	1	N.A.	1000
Linear range (max)	10^5	10^4	Large	10^6	10^4	$\sim 10^3$	2×10^4
Sensitivity at ±1% noise, full-scale	0.002	2×10^{-6}	N.A.	2×10^{-9}	0.01	0.005	0.05
Sensitivity to favorable sample	2×10^{-10} g/ml	1×10^{-7} g/ml	50 cpm ^{14}C/ml	10^{-12} g/ml	10^{-6} g/ml	10^{-11} g/ml	10^{-8} g/ml
Inherent flow sensitivity[b]	No	No	No	Yes	No	No	Yes
Temperature sensitivity	Low	10^{-4}°C	Negligible	1.5%/°C	Low	Low	2%/°C

[a] N.A., not available.
[b] Because of sensitivity to temperature changes, some detectors *appear* to be flow sensitive.

Table 6.3 Characteristics for Proper Selection of the Detector

Required performance criteria of:
Sensitivity
Specificity
Detectability
Linearity
Repeatability
Dependability
A sufficiently small detector cell dead volume such as not to cause excessive peak broadening
It should not be affected by its laboratory location (drafts and room temperature fluctuations)
Insensitive to mobile phase temperature and flow variations
Nondestructive where fraction collection may be considered
Rugged and easy to use and maintain

Electronic Noise

Inherent with most detectors is some form of electronic noise. This can be caused by line voltage fluctuations, high frequency noise within the detector's electronic circuitry, and radio-frequency interferences from outside sources. This noise is usually seldom observed except at high detector sensitivity. Most electronic noise on a straight baseline, for example, will simply cause the recorder pen to vibrate back and forth, producing a baseline with continuous small positive and negative spikes. This is commonly referred to as "grass" and can be demonstrated by turning the recorder gain way up. If it is high frequency noise such as >1 cycle/5 sec (25), there will be little effect on the detector output since the noise is high in frequency and low in amplitude. Electrical noise can also be caused by other instruments on the same circuit such as a heating bath switching on and off.

Short Term Noise

Short term noise also appears as a fuzzy baseline, but it follows the recorder tracing along the eluting peaks. Noise purely electronic in nature is also referred to as short term although it may not be the only cause of what is commonly called short term noise. Short term noise is usually in the range of 1 cycle/5 sec to 1 cycle/5 min (25).

Long Term Noise

This noise appears as slow peaks and valleys along the baseline. It is primarily caused by system variations such as gas in the detector cell, flow rate and pressure variations, leaks, and temperature variations. This noise makes it difficult to draw a straight continuous baseline along the entire chromatogram.

6.2.2 Drift

Drift is a detector baseline that may be straight but continuously moves up or down and off the recorder chart paper. This requires continuous rezeroing and makes any analysis nearly impossible to complete depending on the severity of the drift. This type of drift is also referred to as long term drift, as opposed to short term drift which is a deviating baseline, but not necessarily in only one direction, for long periods of time. Drift can be caused by leaks, temperature variation, change in the mobile phase concentration, a bleeding liquid phase from a column, or a gas bubble in the detector cell.

6.2.3 Sensitivity

Sensitivity or the minimum detectable quantity (MDQ) is commonly defined as the amount of solute detected that is twice the amplitude of the baseline noise. It is also referred to as the MDQ with a signal-to-noise ratio of 2:1.

6.2.4 Linearity

Linearity is that concentration range over which a solute exhibits a constant slope of response to concentration. For every increment of increase in concentration there should be an equal and proportional increase in the response. Most users plot detector response versus concentration on graph paper. They plot three or more points and draw a straight line which usually passes through the origin. Remember, the MDQ is the lower limit of the linearity range and is not necessarily zero. Many plots may appear linear but actually curve ever so slowly. The best fit line is not always a good rough estimation of linearity. The user should always calculate a point by point slope and determine if the differences are within reasonable experimental error before deciding on the linear range.

6.3 Refractive Index Detectors

There are two common types of refractive index detectors, the deflection refractometer and Fresenel. Both detectors are universal in scope and can measure solutes whose refractive index is different from the mobile phase. Table 6.4 includes refractive indices for many commonly used HPLC solvents. A more extensive list covering over 900 organic solvents has been published by Schneider (27).

6.3.1 Variation in Refractometer Detector Response

Both refractometers are very sensitive to changes in temperature, flow rate, back pressure, mobile phase concentration, and pump pulsation.

Temperature change is a common problem with these detectors. Many solvents used in HPLC have a temperature coefficient of about 3×10^{-4} refractive index units/°C, thus requiring temperature control of better than a hundredth or a thousandth of a degree according to the detector sensitivity required (28). Depending on the extent and the direction of change (warmer or cooler) and on the duration of the variation, the baseline will drift up or down. The baseline can cycle if the detector is in line with a fan or blower from another instrument or the laboratory ventilation system. This can be reduced or eliminated by use of several preventive measures.

1. Use the water jacketed lines provided with the detector to control the cell temperature. This is a must for good routine work. (Be sure that the water circulation pump and heater are not on the same electrical circuit to avoid baseline pulses caused by the on/off action of the circulator.)
2. Cover the column from drafts by means of an oven, tubing, towel, and so on (see Section 5.3.3).
3. Cover the inlet line or lines to the detector with tubing or cover with a towel as in 2 above.
4. Never place the detector by an unshaded window.
5. Never place the detector by a door, window, or laboratory hood where drafts are present.
6. Check the mobile phase for temperature changes.

Flow rate and back pressure changes can cause long term noise. These

Table 6.4 Solvent Properties of Some Common HPLC Solvents (J. Calais, *Altex Chromatogram*, Vol. 2, No. 1, Altex Scientific, Inc., Berkeley, CA, 1978.)

Solvent	ϵ^{0} [1]	δ [2]	Viscosity[3]	UV Cutoff[4]	R.I.[5]	B.P.[6]
Acetic Acid	1.0	13.01	1.31(15)		1.372	117.9
Acetone	0.56	9.62	0.30(25)	330	1.359	56.3
Acetonitrile	0.65	12.11	0.34(25)	190	1.344	81.6
Benzene	0.32	9.16	0.65	278	1.501	80.1
1 - Butanol		11.60	2.95	215	1.399	117.7
2 - Butanol		11.08	4.21	260	1.397	99.6
n - Butylacetate		8.69	0.73	254	1.394	126.1
Butylchloride		8.37	0.47(15)	220	1.402	78.4
Carbon tetrachloride	0.18	8.55	0.97	263	1.460	76.8
Chlorobenzene	0.30	9.67	0.80	287	1.525	131.7
Chloroform	0.40	9.16	0.58	245	1.446	61.2
Cyclohexane	0.04	8.19	0.98	200	1.426	80.7
Cyclopentane	0.05	8.10	0.44	200	1.406	49.3
0-Dichlorobenzene		10.04	1.32(25)	295	1.551	180.5
Dimethylacetamide			2.14(20-4)	268	1.438	166.1
Dimethylformamide		11.79	0.92	268	1.430	153
Dimethyl sulfoxide	0.62	12.8[7]	2.20	286	1.478	189.0
Dioxane	0.56	10.13	1.44(15)	215	1.422	101.3
2-Ethoxyethanol			2.05	210	1.408	135.6
Ethyl acetate	0.58	8.91	0.46	256	1.372	77.1
Ethylene dichloride	0.49	9.86	0.79	228	1.445	83.5
Ethyl ether	0.38	7.53	0.24	218	1.352	34.6
Glyme			0.46(25)	220	1.380	93.0
Heptane	0.01	7.50	0.42	200	1.388	98.4
Hexadecane			3.34	200	1.434	287
Hexane	0.01	7.27	0.31	200	1.375	68.7
Isobutyl alcohol		11.24	4.70(15)	220	1.396	107.7
Isohexanes						59-63
Methanol	0.95	14.50	0.55	205	1.328	64.7
2-Methoxyethanol		11.68	1.72	210	1.402	124.6
2-Methoxyethyl acetate				254	1.402	144.5
Methylene chloride	0.42	9.88	0.45(15)	233	1.424	39.8
Methylethylketone	0.51	9.45	0.42(15)	329	1.379	79.6
Methylisocimylketone		8.65		330	1.406	- 144
Methylisobutylketone	0.43	8.58	0.54(25)	334	1.396	116.5
N-Methylpyrrolidone			1.67(25)	285	1.488	202
Nonane		7.64	0.72	200	1.405	150.8
Pentane	0.00	7.02	0.24	200	1.357	36.1
Petroleum ether	0.01		0.30	226		30-60
β-Phenethylamine				285	1.529(25)	197-198
1 - Propanol	0.82	12.18	2.26	210	1.386	97.2
2 - Propanol	0.82	11.44	2.86(15)	205	1.377	82.3
Propylene carbonate		13.3[7]			1.419	240
Pyridine	0.71	10.62	0.95	330	1.510	115.3
Tetrachloroethylene		9.3	0.93(15)	295	1.506	121.2
Tetrahydrofuran	0.45	9.1[7]	0.55	212	1.407	66
Tetramethyl urea				265	1.449(25)	175.2
Toluene	0.29	8.93	0.59	284	1.497	110.6
Trichloroethylene		9.16	0.57	273	1.477	87.2
1,2,2-Trichloro-1,2,2-trifluoroethane			0.71	231	1.356(26)	47.6
2,2,4-Trimethylpentane	0.01	6.86	0.50	215	1.391	99.2
Water	Large	23.53	1.00	<190	1.333	100.0
Xylene	0.26	9.06	0.81	288	1.505	144.4

Notes

[1] ϵ^{0}-solvent strength parameter with alumina as adsorbent.

[2] δ-solubility parameter calculated by Hoy from vapor pressure data.

[3] Viscosity in centipoise (cp) at 20° C unless otherwise noted, e.g., 0.30(25) would be viscosity at 25° C.

[4] UV Cutoff - a maximum value for high purity solvents determined at B & J labs. The UV cutoff is defined as the wavelength in nm at which the absorbance of the solvent measured in a 1 cm cell against a distilled, deionized water reference is 1.0(10% transmittance).

[5] R.I. - refractive index at 20° C unless otherwise noted, e.g., 1.529(25) would be refractive index at 25° C.

[6] B.P. - boiling point at a pressure of 760 mm of mercury.

[7] solubility parameter from Snyder.

References

K. L. Hoy, *J. Paint Tech.*, 42, 541 (1970).

J. Przybylek, "BJ - 25h", 1978.

L. R. Snyder and J. J. Kirkland, *Introduction to Modern Liquid Chromatography*, Wiley Interscience, New York, 1974.

J. A. Riddick and W. B. Bunger, *Organic Solvents*, 3rd Edition, In *Techniques of Chemistry*, A Weissberger (ed.), Vol. 2, Wiley Interscience, New York, 1970.

R. C. Weast (ed.), *Handbook of Chemistry and Physics*, 51st Edition, Chemical Rubber Co., Cleveland, 1970.

L. R. Snyder, *Principles of Adsorption Chromatography*, Marcel Dekker, ed, New York, 1968.

P. A. Bristow, *LC in Practice*, hetp Macclesfield, UK, 1976.

variations cause changes in the mobile phase refractive index which are picked up by the detector. This is particularly noticeable if a static reference is used, which is very common with the deflection refractometer. This noise may also be indicative of a system leak around a fitting. It is important to identify any leaks when doing quantitative analysis, since these leaks can cause random losses of solute and distort the data obtained. Pulses are observed as spikes and/or cycles in the baseline. They are caused by variations in the pump stroke itself, the pump reservoir refill, or gas bubbles in the pump head (see Section 3.4.2). This response may be one way the user can identify gas in the pump head. Follow the techniques described in Chapter 3 to eliminate this effect.

Mobile phase concentration changes will also cause corresponding changes in the refractive index when more than a single solvent makes up the carrier. This will usually be observed as drift. Some users will take samples of the mobile phase at startup and later during use for analysis by GC to determine if significant concentration changes have occurred.

Some of these effects can be minimized by using a dynamic reference with a split mobile phase stream and identical column to produce a matched system to the sample side. However, this will not eliminate these effects completely. A perfectly matched system is rarely attainable. Gradient elution techniques can be used with refractive index, but usually at lower sensitivities. A dynamic reference is necessary, and quantitative studies are not recommended. Finally the user must be prepared for positive and negative peaks. Both can be a result of a real solute whose refractive index is greater or smaller than that of the mobile phase.

6.3.2 Deflection Refractometer

The deflection refractometer detects differences in the refractive index between the sample and the reference cells by measuring the degree of change in the angle of the incident light beam. This is schematically diagramed in Figure 6.1 (1). As the light beam is deflected across the detector, an electrical signal is generated, and the degree of deflection becomes proportional to the concentration. The zero adjust allows the user to move the beam to zero the signal.

This refractometer allows for a static reference cell. The user must remember to purge the reference cell routinely and to avoid immiscible pools or the wrong solvent in the reference side as mobile phases are changed.

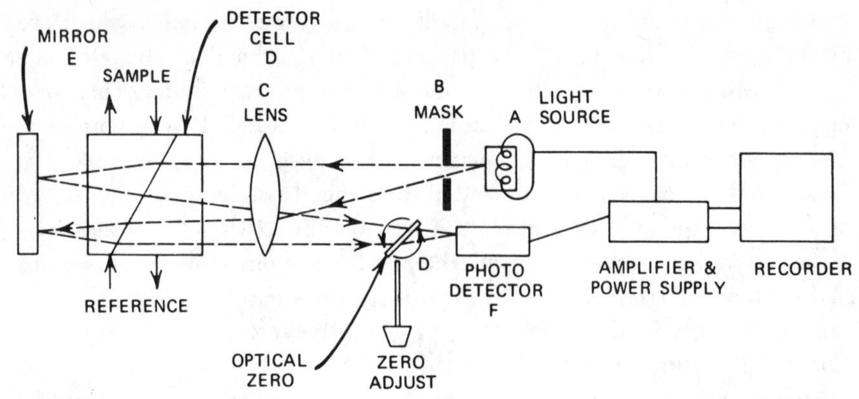

Figure 6.1 Schematic of the deflection refractometer (reprinted with the permission of the publisher).

Cleaning the cell (see Section 6.5) is easy on this detector as well as removing it from the detector. Always remember to check for leaks, particularly after having removed the cell or when cracking any fitting. A leak in this detector can ruin it and may result in having to send it back to the manufacturer. This system is housed in a deep, black painted reservoir for easy access, thermal control, and optical control. With the proper solvent and a leak it is possible to fill the reservoir with the solvent and dissolve the paint off the housing walls and redeposit it on the optics. When checking for leaks it is a good idea to check this detector.

Good CPMA can save the user time and trouble later. The following routine procedures are recommended, both as a measure of the detector's performance but also as a troubleshooting guide.

1 Perform a daily sensitivity check.

a. Turn the attenuation to 16x.

b. Turn the course zero. A ¾ turn to full turn should give 100% full scale deflection.

c. If it takes two turns of the course zero, the sensitivity is only half of what it should be.

d. Record and date these results.

2 Perform these checks weekly or at startup after a period of inactivity.

a. Electronic check.

(1) Turn the attenuator to zero.

(2) If it operates correctly, the recorder should draw a straight line.
(3) Record and date these results.

b. Optics check.
(1) Turn the attenuator to the desired operating range.
(2) With the mobile phase in the cell, turn the pump off.
(3) The recorder should draw a straight line.
(4) Record and date these results.

When problems occur with this detector the above procedures should be tried first. All three tests are performed without flow. They can indicate if the problems are strictly electrical or optical in nature such as a dirty cell, a misaligned optical bench, or a flickering or burned out source light. If these tests are negative, then turn the pump on and look for leaks, old mobile phase in the reference side, or bubbles moving through the detector cell. These steps should be followed before calling a service representative.

To determine if air bubbles are in the detector cell, remove the cover plate and place a yellow piece of paper in front of the photo detector. Bubbles will appear as either stationary or moving dark spots on the light image. Methods to remove gas bubbles are discussed in Section 6.6.2.

Linearity of the detector is checked by plotting the percent of full scale recorder pen deflection versus the number of turns of the optical zero at the attenuation desired. From this plot the degree of nonlinearity can be determined. The detector's operations manual will describe this technique in detail.

6.3.3 The Fresnel Refractometer

The Fresnel refractometer measures refractive index changes between the mobile phase and the reference by monitoring the light reflected at the cell-mobile phase interface. The amount of light reflected is inversely proportional to the refractive index. This detector is diagramed in Figure 6.2 (1).

The detector cell is made from a scored steel base plate, a Teflon spacer, and a prism. A convenient feature of most Fresnel refractometers is the ability to visually observe the detector cell through a viewing window. This allows the user to observe the cell for dirt particles, dirt buildup, air bubbles, and leaks. The cell volume can be changed by simply changing the thickness of the Teflon spacer between the prism and the base plate.

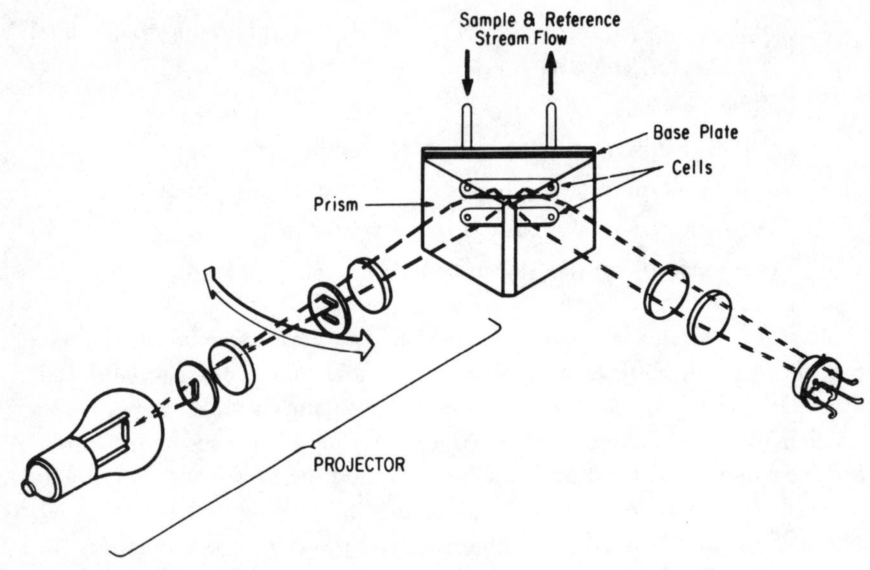

Figure 6.2 Schematic of the Fresnel refractometer (reprinted with permission of the publisher).

This design allows for smaller cell volumes and less drift. This is partly facilitated by the normal operation of the detector, requiring a dynamic reference. However, this detector has a larger glass-mobile phase interface for dirt to adhere to and accumulate. It also is more sensitive to flow and pressure variations, bubbles, dirt, and it has a smaller linear dynamic range. However, it is less expensive than the deflection model.

The detector cell is easy to get at by simply removing the optics unit cover. As with any light sensitive monitor, always turn the power off to avoid overloading the detector and the recorder. Two hexagonal screws hold the prism in place with a narrow support bar. The user needs to be *very careful* on reassembly. With all the parts dry and in place, the prism is carefully tightened onto the base plate. Not enough torque will allow the cell to leak, and too much will crack the prism and/or deform the base plate. This cell is not always easy to leak seal on initial tightening. It is therefore recommended that the user have a spare prism and Teflon spacers available before dismantling. Also be sure to have the correct prism. The standard prism is used with mobile phases having refractive indices between 1.31 and 1.45, while the high range prism operates between 1.40 and 1.55. Table 6.5 lists some HPLC solvent-prism guidelines.

No-flow electronic and optics checks can be performed on this detector similar to the deflection model. In addition it has several other adjustments. A track adjustment is provided to compensate for differences in the electronics to ensure that the photodiode response will be the same for the same changes in either the sample or the reference cells. As outlined in the operations manual, this adjustment should be carried out each week, after a change in mobile phase, or after changing the detector cell. The panel meter provides an indication of the total current flowing in the measuring bridge so that the user can optimize the angle of incidence for maximum sensitivity. For quantitative analysis the meter should read between 0.4 and 0.6 mA. When developing a quantitative procedure, the method described by the manufacturer using the analyte solution should be followed. This permits determining the optimum meter setting for the best response to carry out the study. Detector calibration is similarly described as for the deflection refractometer using a sucrose solution.

To avoid problems with this detector, the following points should be observed:

1 Use the detector heat exchanger and a circulation bath to control the detector temperature (see Section 6.7).

Table 6.5 HPLC Solvent-Prism Guidelines

Solvent	Refractive Index	Temperature (°C)	Prism Type
Acetone	1.358	20	Low
Benzene	1.501	20	High
Carbon tetrachloride	1.463	15	High
Chloroform	1.446	18	High
Ethyl alcohol	1.362	20	Low
Hexane	1.375	20	Low
Tetrahydrofuran	1.404	25	Low
Toluene	1.496	20	High
Water	1.333	20	Low
Methanol	1.326	25	Low
Acetonitrile	1.342	25	Low
Heptane	1.385	25	Low
Methylene chloride	1.424	20	High

2 Make sure that the reference system is matched with the sample side for uniform back pressure and flow.
3 Protect the HPLC system from drafts.
4 Check that there are no leaks in the system.
5 When using a constant volume pump, minimize the pump pulsations with appropriate pulse dampeners.

6.4 ULTRAVIOLET-VISIBLE DETECTORS

The UV-visible detector is probably the most commonly used detector with HPLC systems today. The high incidence of UV absorbing compounds and the availability of quality optic systems thanks to the expertise developed in the field of spectroscopy have made this detector so useful. An optical schematic of a typical variable UV detector is shown in Figure 6.3 (29) along with an expanded view of a UV liquid cell from a single beam detector illustrated in Figure 6.4 (30).

These detectors originally had a single 254 nm mercury lamp source. Fortunately most compounds capable of absorbing UV light give a response with this source. Later detectors with 280 nm capability and subsequently continuously variable UV-visible detectors from 200 to 600 nm have arrived on the market. This gives the user the capability of studying an analyte at or near its wavelength maximum. It is therefore important, when selecting a detector wavelength, to determine the UV-visible wavelength maxima of the compounds to be studied. It is also important that the mobile phase have an optical window in the range of interest. The information in Figure 6.5 can be used for solvent selection windows. Difficulty is usually encountered with studies below 220 nm. There are few solvents available with UV cutoffs below 220 nm, and those that are used must be pure, including water.

Compared to refractometers, UV detectors are relatively insensitive to flow rate and back pressure changes. However, the user must be aware that pressure and flow changes can cause baseline shifts. This occurs because of refractive index changes in the cell. Usually these effects will not cause the user great problems. However, with very sensitive settings, these shifts or drifts can cause problems. The changes in refractive index cause the incident light to be deflected and primarily absorbed on the cell walls. Against the reference side, this appears as absorbance, even without a sample present.

This problem can be prevented or minimized in several ways. Similarly as

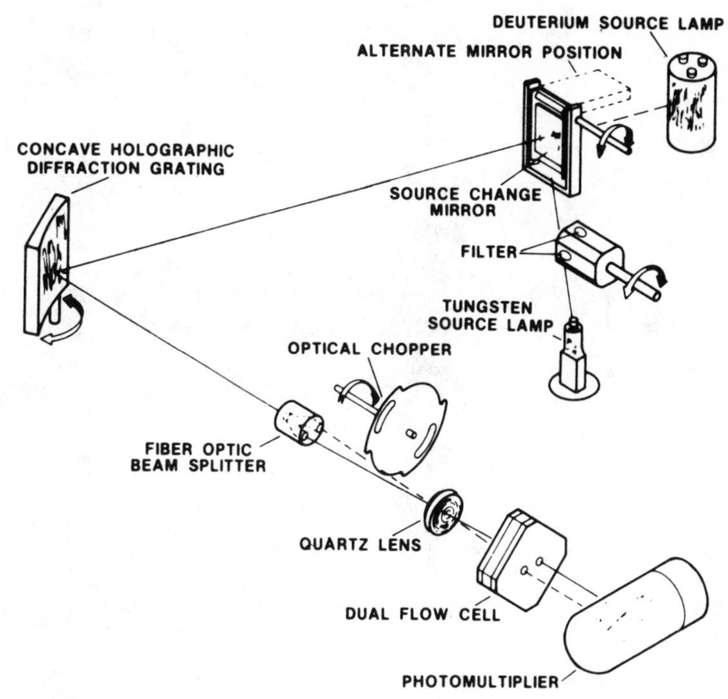

Figure 6.3 Optics diagram of a variable wavelength UV detector (reprinted with permission from *American Laboratory*, Vol. 8, No. 3, 1976. Copyright 1976 by International Scientific Communications, Inc.).

with the refractometer, protect the detector from temperature fluctuations such as drafts. The inlet lines can be covered with insulating tubing of wider internal diameter. Some detectors have cooling fans which draw or blow air over the inlet lines. This draft must be deflected or the lines shielded. Carefully check the mobile phase for temperature and concentration changes. Obviously any increase in back pressure, such as a restriction or a column head starting to clog, will affect the detector. This can be determined by monitoring the flow rate and pressure. Another method has involved a change in the detector cell design. This newer design (Figure 6.6) utilizes a tapered cell which allows the bent incident light to reach the detector sensor without being absorbed on the cell walls. This also compensates for the natural bend in the light stream as a result of the glass-solvent interface, further increasing this detector's sensitivity.

UV detectors are also susceptible to negative peaks. Any non-UV absorb-

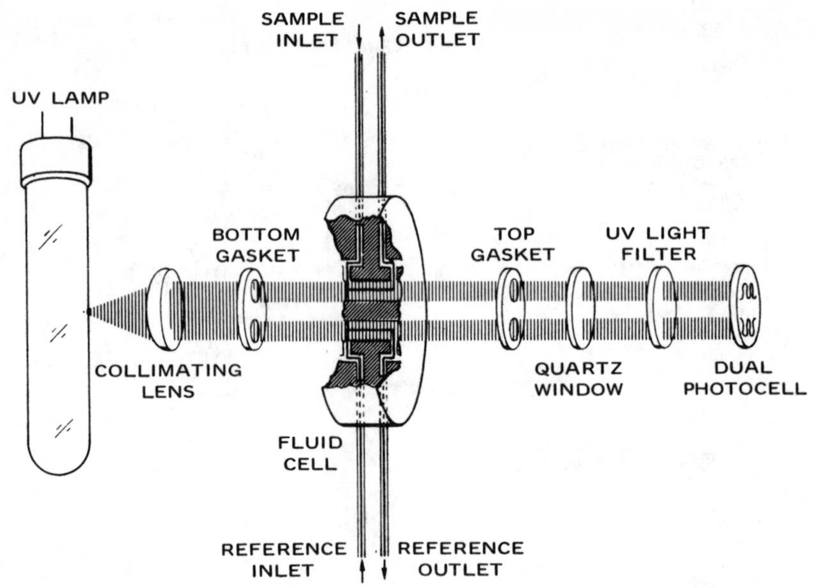

Figure 6.4 Schematic view of a single beam UV detector cell (courtesy of Varian Instruments Group).

ing solute passing through the detector can change the refractive index in the sample cell. This change can allow more light through producing a negative peak on the recorder tracing. Flat peaks can also be produced when the detector electronics become saturated. This happens when the sample concentration becomes too high and the detector reaches nonlinearity. These peaks usually top off at about 60–70% of full scale deflection and usually at attenuations greater than 1.28 AUFS.

Detector response, on the other hand, can drop due to dirt in the detector cell or because of a source lamp going bad. Cell cleaning is described in Section 6.5. To determine whether the source lamp is bad, several procedures are available. Some detectors have digital meter readouts which are used to determine whether enough light is reaching the photocell. These are easy to use and should become part of the user's routine maintenance. Some detectors have panel meters which can help in a similar way. Both techniques tell the user when radiation is not getting through, which can be due to a bad source lamp or to dirt or bubbles. The third approach is to change bulbs. This requires extreme caution. The user must never look directly at a UV source lamp. Any lamp changing should be done with the power off. A user may wish to examine a lamp with the power on. A mercury

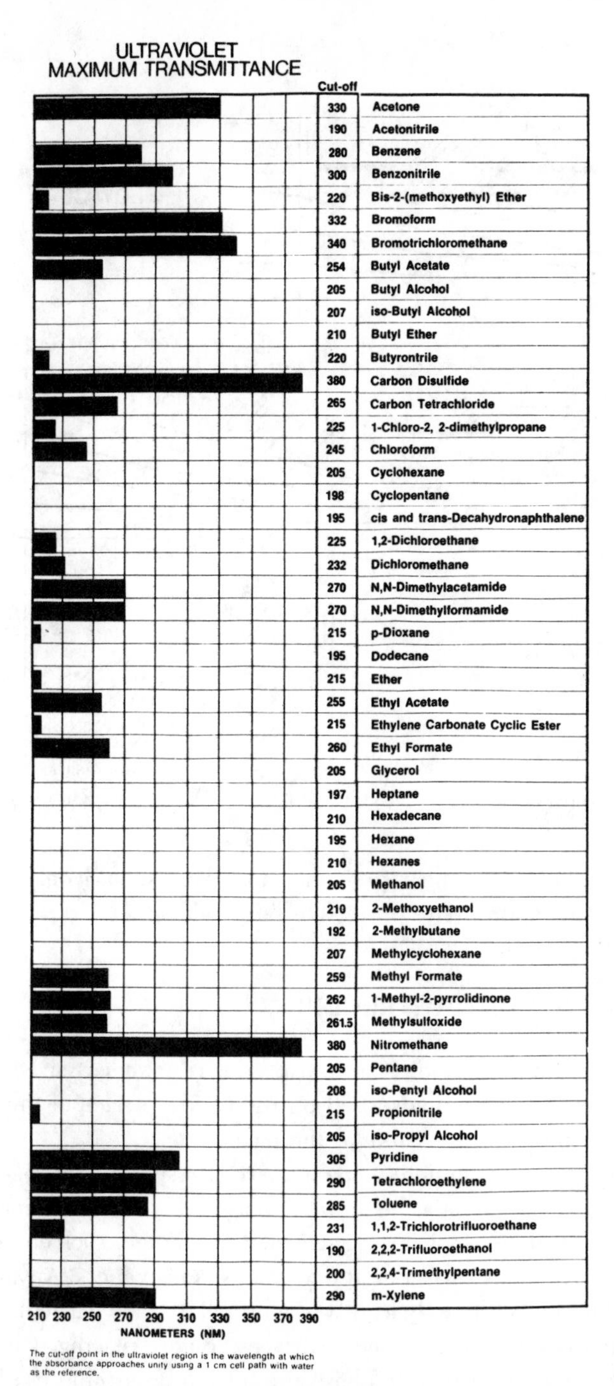

Figure 6.5 Ultraviolet maximum transmittance for some solvents (reprinted with permission of MCB Manufacturing Chemists, Inc., Associate E. Merck, Darmstadt, Germany).

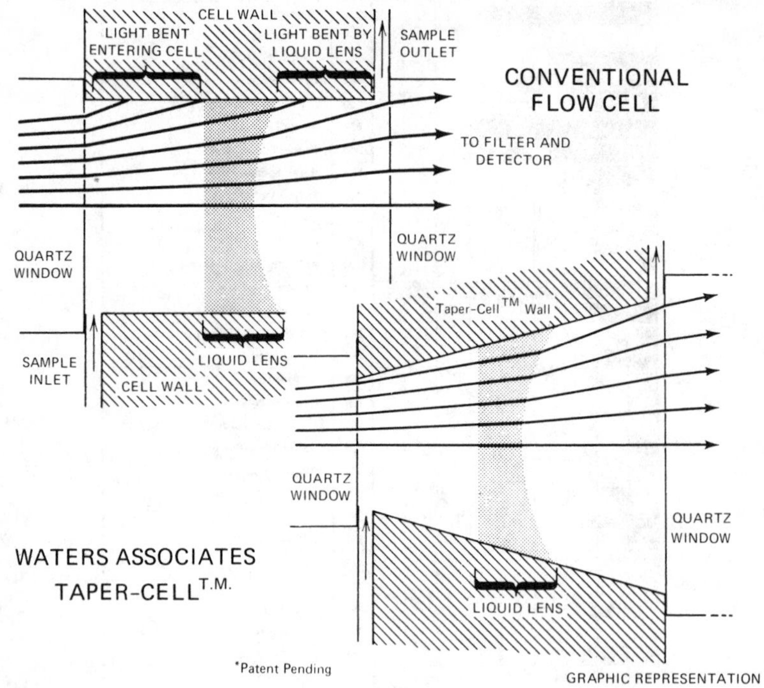

Figure 6.6 Tapered UV detector cell (courtesy of Waters Associates, Inc.).

lamp that is bad will appear pink in color. This should only be viewed indirectly, using appropriate protective safety glasses.

6.5 CLEANING DETECTOR CELLS

Detector cells can become dirty after use. This is not necessarily particulate matter, but contamination adsorbed on the inside surface of the cell walls which may or may not be visible to the naked eye.

Some cells can be completely taken apart. These are held together with screws and O-rings. Usually the entire cell assembly can be dismantled and the cell surfaces and/or lens directly cleaned in a suitable solvent and dried. Never scratch the lens surface. Pat dry and use lens papers. Always be sure to reassemble the cell exactly as it was assembled. Take particular notice of any direction of concave or convex surfaces. Before placing it back in the instrument, try passing mobile phase through to determine if a leak proof seal has been made.

For detector cells that cannot be dismantled, it is possible to flush clean them. This can be done with the following procedure. Wear appropriate safety gloves, apparel and eye protection, and carry this out in a well ventilated area.

1 Disconnect the detector from the column.
2 Flush acetone through the cell with a 50 ml glass syringe with a cannula adapter and appropriate Swagelok fittings.
3 Flush 6 *N* HNO_3 through the cell.
4 Rinse with plenty of tap water (see Section 3.4.2).
5 Rinse with distilled water.
6 Rinse with acetone and air dry.
7 Resume flow with the mobile phase.

This technique can be done in either direction through the cell, and, if desired, the nitric acid or other suitable solvent can be allowed to remain in the cell for short periods of time. Usually this technique removes most contaminates.

6.6 GAS BUBBLES

The biggest problem users have are gas (air) bubbles trapped or passing through the detector cell. This occurs mainly for two reasons: leaks in the HPLC system fittings and/or degassing of the mobile phase as it leaves the column and enters the detector. Leaks can cause air to be drawn into the mobile phase system. Degassing in the detector system is caused by air dissolved in the mobile phase being released as the solvent stream pressure is dropped to atmospheric conditions. Air that is dissolved in the mobile phase because of the high pressure of the system and/or air already dissolved in the solvent can become dislodged by the agitation and action of the sudden pressure drop. The problem is worse when a bubble is lodged in the cell. Techniques for preventing gas bubbles and removing them are described below.

6.6.1 Preventing Gas Bubble Formation

Gas bubbles develop in detector cells for one of two major reasons: pumping bubbles through the HPLC system and solvent degassing in the detector cell. Both can be prevented by practicing good preventative techniques and using good CPMA.

Bubbles being pumped through the HPLC system are the lesser of the two causes. They can be prevented by the following procedures:

1 Make sure that there are no leaks in the system.
2 Expel any air from the pumping system and prevent pump starvation (lack of solvent in the pump head).
3 Avoid using very volatile solvents like ether and pentane. These solvents can undergo cavitation and volatilize in the system (see Section 3.4.2).
4 Do not stir the mobile phase reservoir so vigorously to cause air bubbles to be fed into the pump (see Section 2.4).

The primary cause of bubble formation in the detector cell is solvent degassing. Under HPLC pressure conditions gases become more soluble in solvents and as this pressure is quickly dropped to atmospheric conditions, such as near the column-detector interface, much of this dissolved gas is released and can pass through or lodge in the detector cell as a gas bubble. The foremost preventative measure is to degas the mobile phase prior to use (see Section 2.3.1). However, even prior solvent degassing may not completely prevent some degassing from occurring in the detector cell.

After solvent degassing, the best preventative measure is to maintain the detector cell above atmospheric conditions. This minimizes the column-detector pressure drop and allows for a high gas solubility in the mobile phase passing through the detector. Several techniques that will accomplish this are listed below. *Caution* is advised, however, with any of these methods. Never exceed the pressure limit of the detector cell. The average detector cell (refractometer, UV-visible, fluorimeter) has a pressure limit between 200 and 500 psi. These limits vary among manufacturers, so check the specifications for the detector being used.

1 Connect 10 or more feet of 0.009 ID stainless steel or Teflon tubing to the detector outlet. This acts as a flow restrictor and will maintain the detector cell pressure above atmospheric conditions. Maintaining the tubing outlet above the detector outlet will also help.
2 Attach ⅛ inch stainless steel tee to the outlet of the detector via appropriate tubing. Perpendicularly to the tee connect a 100–200 psig

pressure gauge. To the outlet of the tee (directly in line with the detector line) attach a piece of Teflon tubing which passes through a male-female Swagelok fitting with a Teflon reducing ferrule. Using suitable end wrenches, gradually tighten down on the outlet fitting constricting the Teflon tubing. This will restrict the detector flow and increase the detector pressure. With the pressure gauge the pressure increase can be monitored to safe detector cell levels. Usually 60–100 psig works satisfactorily.

6.6.2 Eliminating Gas Bubbles

Gas bubbles in the detector will cause baseline spikes, ripples, and long term noise. With many detectors the user can visually examine the detector cell and see if a bubble is present. (*Be careful* not to view a cell with UV radiation passing through to the eye. Always wear protective safety glasses, turn the UV source off, adjust the light to the visible region, or view the cell path indirectly, such as with a piece of white paper, and look for a bubble shadow.)

If gas bubbles are suspected, two quick procedures are recommended for bubble elimination.

1 Hold a piece of rubber septum over the detector outlet for brief periods of time (~30 sec). This will increase the cell pressure, dissolving some or all of the gas bubble. If a constant volume pump is used, the pressure increase can be seen with the pump pressure gauge or meter. It is important not to let the pressure build too high to break the cell. Practice and experience will allow the user to properly estimate the timing of this technique. These repeated pressure surges will reduce and eventually dissolve and remove the bubble. If the bubble cannot be directly viewed, improvements can be observed by monitoring the detector baseline. Many chromatographers perform this technique with their thumb or finger. This is not recommended with organic or organic containing mobile phases since these solvents can be absorbed through the skin.

2 The outlet of the detector can be connected to a syringe (31). The mobile phase has to push against the syringe plunger, increasing the back pressure in the detector. If a large enough syringe is used, it can be left on sufficiently long to dispel the bubble.

The methods described in Section 6.6.1 can also be used to eliminate bubbles. Bubbles in aqueous systems are usually more difficult to remove. If a bubble cannot be removed by any of the techniques described here, fill the detector cell with methanol and repeat any of the above procedures.

6.7 DETECTOR CHECKLIST

The detector output is a useful tool in identifying and solving HPLC system problems. Many problems, causes, and solutions are outlined in Appendix 1. This section should permit identifying a number of problems by a particular detector tracing.

In addition there are many detector CPMA do's and don't's the user should bear in mind. Many of these are listed below.

Do's

1 Always degas the mobile phase solvents before use.
2 Plug the detector in and turn it on.
3 Be sure that the detector and the recorder are on the same millivolt setting.
4 Have the detector polarity in the proper direction to coincide with that of the recorder.
5 Label the detector as to when the source lamp was last changed.
6 Have a list and supply of maintenance spare parts. This should include:

Prisms	Lens
Teflon spacers	Special fittings
Source lamp or bulb	Liquid seals
Fuse(s)	Gaskets

7 Have your electronics specialist examine the electric schematic for the most likely components to go bad and stock these.
8 Use the heat exchangers and controlled temperature circulation baths on refractive index detectors and other detectors where applicable.

Don't's

1 Never look directly at a UV source lamp without eye protection.
2 Do not use Teflon tubing between the column and the detector since Teflon tubing wall thickness varies and is permeable to air.

3 Do not inject samples with air bubbles in the sample stream.
4 Never place the detector in direct sunlight or drafts.
5 Do not connect a detector on the same line as alternating on/off equipment such as heating baths, unless proper line filtering has been arranged.
6 Do not use mobile phase solvents that will interfere with solute detection. Be sure that the appropriate UV or refractive index window is used.

REFERENCES

1 L. R. Snyder and J. J. Kirkland, *Introduction to Modern Liquid Chromatography*, 2nd Edition, Wiley-Interscience, John Wiley & Sons, Inc., New York, 1979.

2 J. Waters, *American Laboratory*, **3**(5) 61–65 (1971).

3 F. W. Karasek, *Research and Development*, **7** 20–24 (1975).

4 F. W. Karasek, *Research and Development*, **12** 23–25 (1971).

5 D. Rodgers, *American Laboratory*, **9**(2) 133–138 (1977).

6 S. H. Byrne, Jr., in *Modern Practice of Liquid Chromatography*, J. J. Kirkland, Ed., Wiley-Interscience, John Wiley & Sons, Inc., New York, 1971.

7 J. C. Steichen, *Journal of Chromatography*, **104**, 39–45 (1975).

8 J. L. DiCesare and T. J. Porro, *Trends in Fluorescence*, **1**(1) (1978), Perkin-Elmer Corporation, Norwalk, CT.

9 P. T. Kissinger, *Analytical Chemistry*, **49**(4) 447A–456A (1977).

10 B. Fleet and C. J. Little, *Journal of Chromatographic Science*, **12**(12) 747–753 (1974).

11 An Infrared Detector for Use with Gel Permeation and Liquid Chromatographs, Application Report: Infrared Methods of Analysis No. 2, Wilks Scientific, 1976.

12 V. Pretorius and J. F. J. VanRensburg, *Journal of Chromatographic Science*, **11**(7) 355–357 (1973).

13 A. T. Chamberlain and J. S. Marlow, *Journal of Chromatographic Science*, **15**(1) 29–31 (1977).

14 P. R. Boshoff, B. J. Hopkins, and V. Pretorius, *Journal of Chromatography*, **126**, 35–41 (1976).

15 W. H. Fadden, D. C. Bradford, D. E. Games, and J. L. Gower, *American Laboratory*, **9**(10) 55–64 (1977).

16 F. E. Brinckman, W. R. Blair, K. L. Jewett, and W. P. Iverson, *Journal of Chromatographic Science*, **15**(11) 493 (1977).

17 D. J. Freed, *Analytical Chemistry*, **47**(1) 186–187 (1975).

18 H. Poppe and J. Kuysten, *Journal of Chromatographic Science*, **10**(4) 16A–18A (1972).

19 J. M. Charlesworth, *Analytical Chemistry*, **50**(11) 1414–1420 (1978).

20 D. M. Fraley, D. Yates, and S. E. Manahan, *Analytical Chemistry*, **51**(13) 2225–2232 (1979).

21 K. R. Hill, *Analytical Chemistry*, **17**(7) 395–400 (1979).

22 R. P. W. Scott, *Liquid Chromatography Detectors*, Elsevier Scientific Publishing Co., New York, 1977.

23 F. W. Karasek, *Research and Development*, 3, 34–38 (1975).

24 S. A. Wise and W. E. May, *Research and Development*, **10**, 54–64 (1977).

25 *Lab Notes* 4, Chromatronix, Inc., Jan. 1971.

26 Solvent Properties, *Altex Chromatogram*, **2**(1) (1978).

27 R. L. Schneider, *Eastman Organic Chemical Bulletin*, **47**(1) 1–12 (1975).

28 R. D. Condon, *Analytical Chemistry*, **41**(4) 107A–113A (1969).

29 D. W. Janzen and D. J. Farley, *American Laboratory*, **8**(3) 43–46 (1976).

30 N. Hadden, F. Bauman, F. MacDonald, M. Munk, R. Stevenson, D. Gere, and F. Zamaroni, *Basic Liquid Chromatography*, Varian Aerograph, Walnut Creek, CA, 1971.

31 E. Roggendorf, *Spectra Physics Chromatography Review*, **3**(2) 8 (1977).

7
Data Handling

7.1 INTRODUCTION

Data handling in HPLC is as important to the success of any experiment or analysis as any other aspect discussed so far. This is not peculiar to just HPLC. As an old saying goes: No job is finished until the paperwork is done. Many tests and experiments have been ruined, lost, and redone because of user error in data identification and handling. This does not only pertain to data interpretation for the analysis of the results, but also to the proper labeling and identification of all maintenance and troubleshooting procedures and work.

A very common mistake users make is to carry out maintenance practices and fail to document the work. This varies from not having any written procedures and work schedules to not logging and/or labeling work carried out on the different HPLC system components. This is neglected first because it is often an area of least interest due to its more administrative nature. Second, the chromatographer often is rushed to complete a task and may feel this takes too much time. Nothing could be further from the truth. The few minutes it takes to label a column before storage or a pump head as to when the seal was changed will save hours and hundreds of dollars in wasted time and purchasing of unneeded replacement hardware because no one could remember what or when it was last done. This is where good CPMA will pay off. It is part of good HPLC technique to properly label and document the analytical results as well as all and any maintenance and troubleshooting practices.

7.2 MAINTENANCE AND TROUBLESHOOTING DOCUMENTATION

Any efficient laboratory will have established written maintenance and troubleshooting policies and procedures. These written policies and practices should be available and understood by all users, including those chromatographers from other departments who share the equipment. This may extend so far as to hold periodic user briefings discussing current policies, practices, and problems. Clear communication will pay big dividends on minimizing misunderstandings and mixups. Although this may not necessarily be considered a maintenance and troubleshooting topic, anything that can cause or prevent HPLC system problems and downtime is part of maintenance and troubleshooting. The key to avoiding personnel caused problems is effective communication and documentation. Several such areas are discussed below.

7.2.1 HPLC Operations Manuals

An HPLC operations manual is a collection of policies and procedures which the user can refer to for guidelines and assistance in performing HPLC analyses and troubleshooting. The extent and depth of information in such a manual will be up to the individual users of the HPLC laboratory. Table 7.1 lists examples of topics a typical HPLC operations manual may contain.

The material covered in each of these subject areas as well as any other topics included should be aimed at the normal operation of the user's specific laboratory. The intent of the manual is not necessarily to rewrite instrument manuals or HPLC texts, but to compile with easy access information that the user should have to meet the operation demands of his/her laboratory and department. In some cases step by step procedures may be needed. These may include instrument manual procedures with additional clarification or simply referencing a procedure in a specific instrument manual.

The manual is intended to save time, for example, by listing whom to contact for problems and services, where the columns are located, what forms should be used for data reporting, and so on. A section like method development can be a guide developed for the class of compounds and types of problems specific to the user. Each laboratory has certain ways in which data are to be measured, calculated, and presented. This too should be included. Much of the manual's content can be left to the needs and ingenuity of the users. It will save a lot of time and prevent problems. At the same time it does not have to be written in one day either. This can be

Table 7.1 Typical Table of Contents of an HPLC Operations Manual

1 Theory and definitions

2 Floor plan and facilities layout of the HPLC laboratory(s)

3 Inventory of all HPLC instruments and system components, including:
- a. Name
- b. Laboratory identification number(s)
- c. Laboratory location
- d. Manufacturer
- e. Suppliers
- f. Date of purchase
- g. Approximate replacement date and cost
- h. Location of instrument manuals
- i. Location of instrument maintenance and troubleshooting logbooks
- j. Technical service representative name, company, address, and telephone number(s)

4 Inventory of all HPLC columns, including:
- a. Name
- b. Laboratory identification number(s)
- c. Laboratory location
- d. Manufacturer
- e. Supplier
- f. Approximate costs
- g. Location of the column logbooks (see Section 5.3.4)

5 Laboratory and equipment startup and shut down procedures

6 Care of equipment and instruments, for example:
- a. Column labeling procedures
- b. Use of instrument diaries
- c. Use of Swagelok fittings and how to cut tubing

7 Mobile phase preparation

8 Sample preparation techniques

9 Column cleaning procedures

10 How to clean and repair a sample injection valve

11 Sample injection techniques

12 Qualitative and quantitative measurements

13 Standard calculations

14 Handling (presentation and storage) of data

15 Method development protocols

16 Purchasing procedures

17 Internal contacts and responsibilities list with addresses and telephone numbers, including:
- a. Supervisor(s) in charge
- b. Laboratory assignments
- c. HPLC manufacturers and suppliers

Table 7.1 (*continued*)

18	Laboratory safety information
19	HPLC reports, forms, and labels, including: a. When they are used b. How to fill them out c. Examples of each
20	List of pertinent national and local HPLC meetings and committees
21	List of reference texts

developed with experience. The task of compiling the manual with scheduled updates should be assigned to one or more of the users.

7.2.2 HPLC Instrument Calibration and Maintenance Manuals

Every analytical laboratory with instrumentation, including HPLC equipment, should have a calibration and instrument manual. The manual should include a table of contents listing every piece of equipment used. Similarily to the operations manual, the inventory list covered in Table 7.1, item 3, should be included again in this manual.

Three sections of information should be included for every piece of HPLC equipment used. The first section should be an information page. This will list who is responsible for conducting the calibration and maintenance procedures, a list of each procedure to be carried out and how often, and a parts list of those items needed for normal maintenance. Secondly, the actual procedures should be included. These may be copies of procedures directly from the instrument manual or procedures specifically developed by the users. The procedures should be tried before being handed out. Finally, the manual should include calibration maintenance report log sheets for every instrument. This should include the items listed in Table 7.2. Again, one of the users should be made responsible for maintaining and updating the manual.

7.2.3 Equipment Labeling

A key to maintenance and troubleshooting is proper labeling and documentation. This can be accomplished with simple maintenance log books and/or labeled tags or tape affixed directly onto the equipment itself. Table 7.3 lists

Table 7.2 Criteria for a Calibration and Maintenance Log

Name of the instrument and identification number(s)
Name and number of the calibration and maintenance procedure(s)
Frequency of the test(s)
Date the test was performed
Identification of the person performing the test
Result of the test(s)
Remarks

many such tags or labels that can be used. They serve as quick reminders to users as to the status of a particular item. The instrument or equipment log book will serve to provide more detailed information. Tagging should be used even if the work was performed by an outside technical service representative.

7.2.4 Troubleshooting Logbook

A troubleshooting logbook should be maintained on each piece of HPLC equipment. This can be included in the calibration and maintenance logbook described in Section 7.2.2, or kept as a separate record. Regardless of the style chosen, an account of the problems and how they were solved should be kept. This can serve as a valuable guide later on when the same problems reoccur. It would be helpful to include who repaired the equipment and what parts were used. A parts list developed from a troubleshooting log can alert users as to the parts most likely to fail and which ones should be stocked.

7.2.5 Laboratory Assignments

One of the most frequent problems where more than one person uses the HPLC equipment is thinking that another person was supposed to stock, repair, or prepare something for the laboratory's operation. As a result, there may not be enough solvent to prepare a mobile phase or enough fittings to repair a leak.

This problem can be avoided by making the necessary assignments among the users. Table 7.4 presents some typical HPLC laboratory tasks. These assignments can be spread out among many users so that no one person has

Table 7.3 Examples of Equipment Maintenance Labeling

1 Solvent preparation equipment. Label solvents used or to be used with specific equipment.

2 Solvent reservoir
   a. Mobile phase solvents
   b. Concentration
   c. Date prepared

3 Solvent delivery systems
   a. Pump heads as to last date seals, plungers, and so on, were changed
   b. Date last used
   c. Solvent last used

4 Sample introduction
   a. Type of septum in the injector
   b. Solvent last used in the injector
   c. Pressure limit of the injector
   d. Date of the last rotor change

5 Columns
   a. Inlet/outlet
   b. Packing material
   c. Solvent last used
   d. Storage solvent if different from c
   e. Date last used

6 Detector. Date source lamp was changed

---

them all. Also, require schedules and logging sheets for each task to maintain accountability.

Another problem with HPLC laboratories is that equipment, tools, and fittings disappear without a trace. Keeping everything locked is one way to minimize this problem. Equipment and tools can be color coded for easy identification. A user list can also be maintained, whereby any user taking equipment, fittings, and so on, must sign out for them. For the latter program someone must be assigned the responsibility to receive the sheets and alert the proper inventory control.

7.2.6 Manufacturers' Instrument Manuals

Another item that disappears from laboratories is the manufacturers' instrument manuals. These are taken home to study, borrowed by users of the same equipment who lost theirs, borrowed by in-house repair personnel, or borrowed by future buyers.

Table 7.4 Typical HPLC Laboratory Tasks

1	Maintain the fittings, hardware, tools, and spare parts inventory
2	Maintain adequate inventories of HPLC solvents
3	Monitor the inventory, usage, and handling of HPLC columns
4	Monitor the updating of the operations manuals, calibration and maintenance logbook(s), and troubleshooting logbook(s)
5	Maintain the supply of instrument manuals
6	Repair worn sampling valves
7	Pack and quality check columns
8	Perform routine maintenance on pumps, detectors, and recorders

This can be avoided by obtaining a second copy from the manufacturer or making another copy in-house. The originals can be kept together, locked, and the supply controlled and monitored by one of the users. The second copy can be kept with the specific piece of equipment. In this way if the active manual is lost, the original is available for use and for future copies. Again, this is good CPMA which can save users lost time.

7.3 STRIP CHART RECORDER

The most common method of data collection is by using strip chart recorders. These permit permanently recorded data for measurements, storage, and retrieval. They are used with many analytical instruments so that most HPLC users are already familiar with them. However, many problems occur with these recorders that cause data to be lost and experiments to be repeated. Many of the problems and recommended practices for users are discussed below. For a detailed description on how these recorders work, Ewing and Ashworth have provided a comprehensive review (1).

7.3.1 Recorder Problems and Troubleshooting

Recorders, like any analytical instrument, are subject to electrical interferences, user mishandling, or parts wearing out. Each recorder has its own instrument manual which should be consulted for proper operation, maintenance tips, and troubleshooting chart. The most common problems encountered with recorders are seldom serious mechanical or electrical malfunctions, but rather simple user omissions. Many of the problems discussed below may seem ridiculously simple, yet so many users actually encounter them.

1 **No recorder power.** Be sure the recorder is plugged in and turned on.
2 **Peaks are too large or too small.** Be sure the detector and recorder are set for the same millivolt output-input setting. Many recorders offer multiple and variable millivolt settings. Frequently the wrong setting can be on or the variable switch turned on and yet not noticed. Also check the detector attenuation.
3 **Power on, no peaks.** Be sure the detector output is attached to the correct recorder pen channel.
4 **Sticking chart paper.** Some recorders, even when adjusted properly, may not drive the paper smoothly without buckling and/or jamming in the pen. A clamp with a suitable weight can be attached to the end of the chart paper to maintain adequate tension. Check to see that the correct chart paper is used. Not all manufacturers' chart papers are interchangeable since the drive gear sprockets are often aligned differently. Check for slipping gears or clutch.
5 **Dry pen.** There are three common ink pens used: felt tip, ball-point, or fluid ink. All three can run dry if they are left uncapped when not in use. Good CPMA will have the user cap the pen after each day's use. Check the pen reservoir before startup each day to be sure there is enough ink. A dry pen can be rejuvenated with techniques such as the following: (1) Felt pens can be dipped in a suitable solvent (water works with some). (2) Fluid ink pens can be opened with a very thin wire, similar to those used to unplug syringe needles. (3) Ball-point pens can be rubbed vigorously over a hard surface. Be sure to maintain an adequate supply of pens and bulk ink in each color preferred.
6 **Noisy pen.**
   a. Gain is set too high.
   b. The recorder cable grounding/shielding is not connected.
   c. Check the detector leads for proper connection.
   d. Dirty slide wire, carefully clean with a suitable cleaning fluid specified by the manufacturer or use a cotton swab dipped in alcohol (1).
   e. Improper ground.
   f. Damping improperly set.

7 **Pen string breaks.** Consult the instrument manual or call the technical service representative of the recorder.

8 **No chart paper.** No roll of chart paper will last indefinitely. Some have color indicators marking the end of a roll. Whether or not there are roll end indicators, always check the roll before starting a run and be sure the laboratory is adequately supplied.

7.3.2 Data Handling

An HPLC chromatogram, if sufficiently labeled, can identify the peaks, allow qualitative and quantitative measurements to be made, and describe the HPLC conditions used. However, if there is insufficient information available, the chromatogram may be worthless.

Most strip chart recorders will allow the user to write on the chart paper. A complete data summary should be made on the chart paper at the beginning of a day's run, an individual analysis, or for a different set of experimental conditions. The easiest way to ensure that proper information is not omitted, is to use preprinted labels or ink stamps. These can be designed especially for the user or purchased commercially from HPLC supply companies such as Alltech Associates, Applied Science Laboratories, Inc., or Supelco, Inc. Figure 7.1 shows an example of a printed self-adhesive label for HPLC chromatograms (A# is an internal sequential analysis code number.)

Another good practice is to run the chromatogram baseline at least at 10% or 90% of full scale deflection, depending on the direction of pen deflection. This allows for drift that could occur while the user is not attending to the instrument. Usually this will allow enough chart paper for deviations to occur and still be corrected in time. For refractive index detectors a 30–50% baseline may be required, depending on the extent of peaks in both directions.

Chromatographic measurements are made either manually or automatically. Manual measurements are usually made with a ruler to measure retention times, peak heights, and peak widths for quantitative calculations. The manual approach is quite satisfactory for most work, although it is not as convenient as data produced with the aid of electronic integrators, computers, or microprocessors. The key point in manual measurements is to use a quality draftsman ruler with sufficient subdivisions. A good test of the laboratory's rulers is to match them up division for division and compare them. Quality rulers will match. Discrepancies can occur in laboratory to

HIGH PERFORMANCE LIQUID CHROMATOGRAPHY

Date ________ Analyst ____________ Notebook ________ Page(s) ________

Analysis for ____________________ in ____________________

Lot # ______________ A# __________ Submitter ______________

Sample conc. ______________ Sample solvent ______________ Inj. vol. ________

Column ______________ Length __________ I.D. __________ O.D. ________

Phase ____________________ Support ____________________

Mobile Phase and Gradient ______________________________

____

Temps (Reservoir) __________ (Col.) ________ (Det. 1) ________ (Det. 2) ________

Pressure __________ psig. Flow Rate __________ ml/min. Chart Speed __________

Detector 1 ______________ nm __________ Attenuation ______________

Detector 2 ______________ nm __________ Attenuation ______________

Derivatization ______________________________________

Figure 7.1 An HPLC self-adhesive label.

laboratory data comparisons with the best chromatographic technique but using poor quality rulers. The same data can be obtained by using on-line computers and microprocessor controlled detectors. The latter can print the chromatogram with retention data, peak area, peak height, and percent composition. Table 7.5 summarizes many of the quantitative techniques used. Information is available on the application of microprocessors to HPLC (3–5).

Data storage is also important. It is not uncommon for important chromatograms to be misplaced or lost. A storage and retrieval system should be established. Most chromatograms come from rolled chart paper and are not always easy to store. They can be stored by the roll or by chromatogram in ring binders, 9 × 12 inch mailing envelopes, or files. Regardless of the method used, there should be a convenient method of identification and retrieval. One method is to trifold the chromatograms and insert them into large mailing envelopes such as 8½ × 11 or 9 × 12 inches. The important chromatographic data are attached to the front of the envelope, and the envelopes are filed chronologically. An example of such an envelope label is shown in Figure 7.2.

Table 7.5 Quantitative Techniques (2)

1	Peak height measurements
2	Height times width at half-height
3	Triangulation
4	Planimetry
5	Cut and weigh
6	DISC ® integrator
7	Digital integrator
8	Computer
9	Microprocessor

Number ______________
Date ______________
Analyst ______________
Notebook ________ Page ______

HIGH PRESSURE LIQUID CHROMATOGRAPHY REPORT

Objective ______________

Substance ______________

Product - Formulation ______________

____

Instrument used ______________

Column: Liquid phase ______________ Support ______________

Length ______________ Diameter ______________ Material ________

Temperatures: Column ________ C°

Eluant ______________

Flow rate ____________ ml./min. at __________ PSI

Concentration: ________ mg. in ________ ml. of __________

Injection Device ______________ volume mcl. equivalent to __________ mcg.

Derivatization ______________

Detection: Type __________ Range ______________ Attenuation __________

Type __________ Range ______________ Attenuation __________

Retention times __________ __________ __________

Results and Comments: (linearity range, limit of detection, stability, etc.)

Figure 7.2 An HPLC report form.

The importance of proper labeling and storage of HPLC data cannot be stressed enough. Much lost time, misdirected experiments, and ruined columns can be avoided by simply handling the data as carefully as the equipment itself.

7.3.3 Recorder Calibration

Laboratory recorders need to be calibrated like any other analytical instrument. The chromatographer depends (1) on the accuracy of the chart drive to make retention time and retention volume measurements, (2) on the sensitivity and noiseless operation of the recorder for trace analysis and (3) on the accuracy and linearity of the response. Again, the HPLC system is only as good as the weakest component in the system. An example of a strip chart recorder calibration procedure is presented in Appendix 2. However, it can easily be modified for other recorders. Again it is important that regularly scheduled and documented calibrations become part of the user's CPMA.

REFERENCES

1 G. W. Ewing and H. A. Ashworth, *The Laboratory Recorder*, Laboratory Instrumentation and Techniques Series, Plenum Press, New York, 1974.

2 E. L. Johnson and R. Stevenson, *Basic Liquid Chromatography*, Varian Associates, Palo Alto, CA, 1978.

3 R. Fincher and D. Hathway, *American Laboratory*, **11**(2) 65–76 (1979).

4 E. L. Johnson and S. J. Luchette, *American Laboratory*, **10**(9) 51–58 (1978).

5 A. Henshall, *Spectra Physics Chromatography Review*, **9**, 6–7 (1978).

8
Laboratory Safety for HPLC

8.1 INTRODUCTION

The practice of HPLC is carried out primarily in chemistry or analytical instrumentation laboratories. These laboratories are designed for safe and efficient use of their facilities. But no matter how well designed the laboratory is, personal safety is ultimately the responsibility of each analyst as well as management. This requires well informed and trained laboratory users. Since the establishment of the Occupational Safety and Health Administration (OSHA), there has been a definite focus by management and analysts alike on becoming properly informed and trained in safe laboratory practices. This has included training programs by industry both in-house and external, such as the short courses offered by J. T. Baker, by national societies such as the American Chemical Society, and by chemistry departments of colleges and universities.

This chapter discusses elements of good, safe laboratory practice pertinent to the HPLC user. In addition, the chromatographer is referred to other safety sources listed in the bibliography at the end of this chapter. However, no amount of information or training will be of value unless the chromatographer practices safety throughout his/her work. Similarly to CPMA discussed in Chapter 1, the user must develop the same attitude toward safety. The fact that a majority of the user's time is spent with instrumentation is no excuse for neglecting safe laboratory practices. There is no excuse for damaging a lung, losing an eye, or getting a hernia. These and many other lost time accidents are possible when performing HPLC. Many of these potential accidents and methods to avoid them are presented below.

8.2 SOLVENTS

The most hazardous materials with which the user is frequently in contact are the volatile and flammable solvents used for mobile phases, sample preparation, flushing, and cleaning. The volume of these solvents will vary from a few milliliters to many liters at a time. The dangers involve toxic fumes, either inhaled or absorbed through the skin, and fires. A table of recent OSHA concentration limits for gases, including vapors from HPLC solvents, can be obtained from Foxboro Analytical, P.O. Box 449, South Norwalk, CT, 06856. The following list highlights many of the precautions HPLC users should take when working with solvents.

1 Store all flammable solvents in OSHA approved flammable storage cabinets (1). Most HPLC laboratories use and store large volumes of flammable solvents, even if large volume central storage and stock areas are provided. The quantities usually range between 20 and 60 gallons. Since liter quantities are usually prepared daily, large gallon bottles are generally stocked. Portable approved storage cabinets are allowed to hold up to 60 gallons. Storing solvent bottles under laboratory benches, on top of the benches, and in hoods only produces clutter and increases the hazard of breakage and spillage. This is particularly important for HPLC users who store their solvents in glass rather than metal or plastic.

2 Ground any metal drums used to store bulk solvents for future transfer. Grounding drums will prevent static electricity from building up which could spark between the drum and the transfer container and cause an explosion or fire.

3 Be sure any safety storage cans or containers are OSHA approved and compatible with the solvent to avoid contamination.

4 Prepare all mobile phases in a well ventilated, noncluttered laboratory hood. The chromatographer typically prepares liter quantities of mobile phase. Toxic vapors are always present when pouring, mixing, or degassing solvents.

5 Prepare samples in a well ventilated, noncluttered laboratory hood. Although the volumes of solvent may not be very large, the user will start with a stock of mobile phase or some other solvent (see Section 4.2) and spend considerable time preparing the samples. Again the amount of fumes and exposure time can be critical.

6 Cover the mobile phase reservoir. This will not only prevent concentration changes from occurring in the mobile phase which will change the ratio of the mobile phase where more than one solvent is present (see Section 2.4), but it will also introduce vapors into the laboratory. This can easily be handled by using a close fitting stopper with a hole through its center for the feed line to the pump, or aluminum foil can be form fitted over the reservoir opening.

7 Flush and purge pump heads carefully and in a well ventilated area. This usually occurs when a solvent like methanol is used to eliminate an air bubble from the pump (see Section 3.4.2). During this process solvent can be running over and down the face of the pump. One method of minimizing the buildup of fumes is to use an elephant hose or some other suitable form of vacuum or aspiration directly above the pump to withdraw the fumes from the user. Also a drain pan should be used to prevent solvent from dripping on the floor.

8 Keep solvents off the skin. When checking for leaks (see Section 5.3.3) or getting solvents on the hands or other parts of the body, quickly rinse and/or flush these areas with plenty of water. Organic solvents can be absorbed through the skin. Most of the time a user will be unaware of this absorption. However, when cleaning columns with DMSO (see Section 5.5.1) the smallest amount of this solvent on the skin will be absorbed. This can be detected quickly by the taste of DMSO in the mouth, which can last for days. Therefore it is not recommended to use the thumb or finger to increase the back pressure on a detector when eliminating air bubbles (see Section 6.6.2).

9 Cover all HPLC system drain containers. This includes sample injector waste containers. As the effluent leaves the system, it is usually diverted to a beaker or flask. This must be covered with a suitable material such as a stopper or aluminum foil, as described in 6 above. Another technique which adds additional safeguards is to run a line from the detector directly into a container in a laboratory fume hood. Where no hood is immediately available, an elephant

hose or inverted funnel connected to the house vacuum or aspiration can be used.

10 Do not squirt solvents from injection syringes on the floor. This is a common habit of most GC and HPLC users. Although it may only be a few microliters, it will add dangerous fumes to the laboratory air, spot the floor and, if aimed wrong, can injure another person.

11 Do not operate HPLC systems in laboratories or areas where open flames, sparks, or excessive heat may be present. This includes bunsen burners, muffle furnaces, ovens, nonexplosion-proof refrigerators, and gas chromatographs, especially those equipped with flame ionization detectors.

12 Always dispose of waste solvents, effluents, samples, and so on, in suitable waste solvent disposal cans. These should be emptied into bulk waste storage for the entire laboratory and incinerated or picked up by an authorized solvent disposal company. Most organic solvents can be mixed in the same waste containers. The halogenated solvents such as $CHCl_3$ and CH_2Cl_2 should always be stored separate from the other organic wastes. Halogenated solvents will react violently, for example, in the presence of acetone. Solvents like dioxane, ether, and THF can form explosive peroxides. These containers should be kept covered and neutralized with aqueous sodium sulfite or a solution of ferrous salt. Peroxide indicating papers are available from J. T. Baker. These solutions can also be diluted and washed down the drain.

13 Have adequate spill control materials or kits (as obtainable from J. T. Baker) in the HPLC laboratory. The following typical materials can be used for spills: sodium bicarbonate for acids; citric acid, boric acid, or 6 *N* hydrochloric acid for bases; and materials like Zorb-all ® and Petro-sorb ® for organic solvents.

14 Properly label every solvent and sample container.

8.3 PRESSURIZED LIQUIDS

Although accidents from pressurized pumping and injection systems are solvent related, a separate section is necessary to stress the nature of these incidents. In most cases leaks and repairs made will not involve strong streams of solvents squirting across the laboratory. Liquids do not compress very much and do not build up great potential energy to come streaming out at the user. Constant volume pumps pose very little danger. However, constant pressure pumps can be a hazard as they will continue to pump, releasing the solvent with some force when disconnected from the rest of the HPLC system. Care must be used when cracking fittings on the pump side of a column when the HPLC unit is in use. A particularly dangerous situation can develop with the simple gas pressure pump described earlier (see Section 3.5.1). When this pump is opened for refilling, it is essential that the gas pressure be relieved to the atmosphere. If not, once the fittings or column head are loosened, mobile phase will squirt all over the user and the immediate laboratory area since the gas will compress and remain pressurized until released. This can be avoided by placing a pressure relief valve in line between the gas supply and the pump. Preparative systems can cause serious inhalation accidents when releasing the vacuum. A large amount of solvent vapor can be released quickly and can burn the user if care is not used.

Reciprocating constant volume pumps often employ restrictors and/or dampening units between the pump and the sample injector. If the system becomes clogged and blocked, it is possible for the pump to continue pumping until the restrictor or dampener explodes. The break in the system will occur at the weakest section, and if the pump's pressure relief system is set above the pressure limit of these devices, they can blow open. A plug of this type can come from a piece of septum logged in some tubing or an injection valve stuck between ports. Although it cannot prevent all pressure related problems, the user should consider purchasing pumps or system devices with pressure limit relief systems which can minimize system component damage and personal harm. Another hazard for users are septa that no longer retain the mobile phase. Septum lifetime is limited (see Section 4.3.1). After many injections a hole is produced in the septum that can no longer be closed by the elastic action of the septum material. At high enough pressure the mobile phase can stream straight out at the user.

8.4 ELECTRICAL

Nearly every component or accessory of the HPLC system uses electricity. This requires that the HPLC laboratory have enough electrical circuits to handle all the instruments and sufficient grounding. When repairing any piece of equipment be sure that the power is off and the system is preferrably unplugged. This will avoid shocks and physical harm from moving parts. If the power is needed to test the equipment, use insulated tools and caution. Never poke tools such as screwdrivers and awls into circuit boards. Capacitors, especially in power supplies, store charges that can dissipate through the user with a good jolt. Always ground them first. Finally have your qualified electrical engineers or service team examine the electrical schematics for any hazards or difficult repairs.

8.5 PHYSICAL INJURIES

Next to solvent related accidents, physical injuries and accidents are most common. These usually occur through carelessness and when no attention is paid to what is going on.

The first area of concern involves physically moving HPLC equipment. It is very common for laboratories to relocate or to have to move the equipment for maintenance and repair work. What users often forget is that the weight of many HPLC instruments and individual components is particularly deceiving, and usually the equipment is not in a position for the safest leverage for the user's back. Furthermore, many users have not kept up with their physical fitness programs and may easily strain and pull muscles when moving equipment. For example, the Perkin-Elmer pump described earlier (see Section 3.4.1) weighs 80 pounds and is not in an easy to lift position. It is recommended that the user have experienced help when unpacking, setting up, and moving HPLC equipment. The appearance of herculean efforts proves nothing when an injury results. Also, safety shoes should always be worn when this equipment is moved.

Some equipment may have spring loaded parts or heavy doors and lids which can cause an injury. The DuPont model 830, for example, has a heavy oven lid. If the lid is not properly secured when opened as the manufacturer has provided, it can drop on a user's arm, hand, or head, causing an injury. Any compartment on a piece of equipment which opens and closes should be checked periodically for safe operation.

Finger cuts are a very common hazard. They can occur while cutting stainless steel tubing, repairing a broken sapphire pump plunger, being stabbed with a syringe needle, and cleaning up glass from broken solvent bottles, sample vials, and so on. Syringes should always be kept in a locked drawer or cabinet and never be placed in a drawer of miscellaneous items for an unsuspecting user to find while rummaging through the drawer.

8.6 DETECTORS

The major problem with detectors is eye damage from the UV source lamps. The user should never look directly at these lamps and should wear appropriate protective eye covering.

Some detectors still use UV source lamps that produce ozone. Ozone in the laboratory atmosphere can cause respiratory problems. These detectors should be operated with a source vacuum directly above the lamp. A simple technique to accomplish this is to place an inverted funnel over the lamp, connected to a hose and the source of a house vacuum or an aspirator.

8.7 GENERAL RULES

The following general guidelines should always be adhered to:

1. Never work alone.
2. Never eat, drink, or smoke in the laboratory.
3. Always remove jewelry and watches when working with the equipment and when carrying out any maintenance or troubleshooting practices.
4. Always wear OSHA approved safety glasses (2).
5. Wear the proper laboratory clothing and safety shoes.
6. Use the proper tools for any repairs.
7. Keep informed and alert to safety practices and attend regularly scheduled safety meetings and seminars.
8. Know where all the exits and the safety equipment are located.
9. Know the telephone numbers of all emergency personnel for your area, including local police, fire, and hospitals.
10. Always read the directions first before starting any assignment.

8.8 CHEMICAL LABORATORY SAFETY BIBLIOGRAPHY

1 W. Braker, A. L. Mossman, and D. Siegel, *Effects of Exposure to Toxic Gases—First Aid and Medical Treatment*, 2nd ed., Matheson Gas, Lyndhurst, N.J., 1977.

2 *Chemical Reference Manual*, Vol 1, Manufacturing Chemists Association, Norwood, OH, 1973.

3 *Fisher Safety Manual*, Fisher Scientific Co., Pittsburgh, PA, 1974.

4 *Guide for Safety in the Chemical Laboratory*, 2nd ed., Manufacturing Chemists Association, Van Nostrand Reinhold Co., New York, 1972.

5 *Handbook of Hazardous Materials, Technical Guide, Number 7*, American Mutual Insurance Alliance, Chicago, IL, 1974.

6 *Handbook of Laboratory Safety*, N. V. Steere, ed., The Chemical Rubber Co., Cleveland, OH, 1967.

7 Occupation Safety and Health Standards, Federal Register 39(125), pt. II, June 27, 1974.

8 *Safe Handling of Compressed Gases in Laboratory and Plant*, Matheson Gas Products, Joliet, IL, 1974.

9 *Safety in Academic Chemistry Laboratories*, American Chemical Society, Washington, DC, 1974.

10 S. Sichak, *The Laboratory Safety Deskbook*, Chicago Section, American Chemical Society, Chicago, IL, 1979.

11 N. I. Sax, *Dangerous Properties of Industrial Materials*, Reinhold Book Corporation, New York, 1975.

12 *Tentative Standard on Fire Protection for Laboratories Using Chemicals*, NFPA No. 45-T, National Fire Protection Association, Boston, MA, 1974.

13 *Threshold Limit Values for Chemical Substances and Physical Agents in the Workroom Environment with Intended Changes*, 2nd printing, American Conference of Governmental Industrial Hygienists, Cincinnati, OH, 1976.

14 Toxic Substances List, National Institute for Occupational Safety and Health, Rockville, MD, revised annually.

REFERENCES

1 Occupational Safety and Health Standards, Federal Register 39(125), pt. II, Subpart 1910.106, pp. 23616–23617, June 27, 1974.

2 S. Sichak, *The Laboratory Safety Deskbook*, Chicago Section, American Chemical Society, Chicago, IL, 1979.

Symbols and Terms

The following is a brief list of definitions for the symbols and terminology used throughout the text.

AUFS — Absorbance Units Full Scale

α (alpha) — The separation factor; measures the degree to which two peaks are separated. Alpha equals unity for two peaks that overlap completely. Alpha is also expressed as:

$$\alpha = \frac{k'_2}{k'_1}$$

k' — Referred to as the capacity factor, k' is a measure of the sample retention on a column or the amount of solvent required to elute the sample off the column in column volumes.

$$k' = \frac{V_1 - V_0}{V_0}$$

k'_1, k'_2 — The capacity factor for peak 1 and peak 2 respectively.

N — Theoretical plates; a measure of a column's efficiency. The narrower a peak, the higher N will be, and the more efficient the column.

$$N = 16\left(\frac{V}{W}\right)^2$$

V_O The column void volume is the total volume of the column minus the volume occupied by the packing. It includes the interstitial packing volume and the pore volume in the packing. A sample eluting at the void volume is not retained by the column.

V_R The elution volume or amount of solvent (ml) used to elute a sample from a column.

R Resolution is a total measure of separation between peaks at their apices and baselines. $R = 1$ means that two peaks are 98% separated.

W Peak width

Appendix

1
An HPLC Troubleshooting Guide

Problems and variations observed in the detector-recorder output (tracings) can aid the chromatographer in diagnosing system problems. This is discussed in Chapter 6.

The following chart lists many frequently encountered symptoms characteristic of HPLC system problems. Each symptom has corresponding suggestions as to the cause of the problem as well as recommendations for corrective action.

Symptom	Possible Cause	Corrective Action
1 No flow or pressure	**1a** System leak	**1a** Locate and repair
	1b Injection valve improperly positioned	**1b** Check valve for proper rotation
	1c Air in the pump	**1c** Disconnect system at pump outlet and pump at maximum flow rate with MeOH until no more bubbles appear; replace mobile phase
	1d Faulty pressure gauge	**1d** Check for proper flow at the system outlet and insert a different gauge
	1e No mobile phase	**1e** Check solvent reservoir and refill
	1f Solvent delivery system not pumping	**1f** Check that the pump is plugged in and turned on; check the pressure limit switch
	1g Pump starvation	**1g** Check the boiling point of the solvent for potential pump cavitation
2 Pump pressure up but no flow through system	**2a** Particulate matter clogging inlet system or head of column	**2a1** Filter mobile phase and sample **2a2** Check syringe for a barb(s) breaking septa pieces off into the system
	2b Leak in system	**2b1** Check all system fittings and repair **2b2** Check detector cell for leaks
	2c Plugged detector line	**2c** Turn pump off immediately and carefully clean lines and cell

Problem	Cause		Remedy	
	2d	Injection valve improperly positioned	**2d**	Check valve for proper rotation
	2e	Column inlet clogged with dirt accumulation	**2e**	Clean inlet and/or replace column
3 Noisy baseline	**3a**	Air bubbles passing through detector	**3a1**	Degas mobile phase
			3a2	Flush pump check valves clear of air
			3a3	Check all fittings for air leaking into mobile phase stream; look for salt-like deposits and stains near fittings; tighten appropriately
	3b	Column packing passing through detector	**3b**	Check column outlet for proper column plug and screen
	3c	Leak in system	**3c**	Locate and repair
	3d	Pulses from pump	**3d**	Add a pulse dampener and/or restrictor
	3e	Pulse dampener(s) and/or restrictors not properly flushed	**3e**	Disconnect solvent delivery system from injector and purge with suitable solvent(s)
	3f	Bubbles in detector sample or reference cell	**3f**	Check for bubbles entering detector and flush out air
	3g	Dirt in detector	**3g1**	Disconnect detector from system and back flush with suitable solvents
			3g2	Clean detector cell
	3h	Detector source failing	**3h**	Check and replace source

Symptom	Possible Cause	Corrective Action
	3i Temperature effects on detector cell input tubing	**3i1** Insulate inlet tubing **3i2** Move instrument away from drafts and/or direct sunlight
	3j Recorder improperly grounded	**3j** Check recorder and properly ground
	3k Noisy electronics	**3k1** Check appropriately detector and recorder circuits; Consult instrument manual **3k2** At high detector attenuation check source lamp **3k3** Check for dirty or loose electronic contacts; also check for instrument vibration
4 Baseline drifting	**4a** Dirt in detector sample or reference cells	**4a** Flush detector cells with solvent or carefully clean cell
	4b Temperature gradient over the system	**4b1** Check for drafts **4b2** Insulate column and column inlet-outlet lines or use a constant temperature jacket **4b3** Move instrument away from direct sunlight
	4c Contamination bleed in system	**4c1** Check for septum bleed and replace with proper septum **4c2** Check for column bleed: i. Previous sample(s) buildup—wash the column ii. Column-mobile phase incompatibility—replace column or mobile phase

		4c3	Stationary phase bleed (particularly at elevated temperatures); check stationary phase solubility in mobile phase; change mobile phase and column, or add stationary phase to mobile phase, or add a heavily loaded precolumn to the system
		4c4	Uneluted peaks; wash column
4d	System leak	**4d**	Locate and repair
4e	Bubble trapped in detector sample or reference cell	**4e1**	Flush out cell
		4e2	Degas mobile phase
		4e3	Locate and repair any system leaks
		4e4	Add suitable back pressure to detector outlet
4f	Solvent immiscibility or immiscible pools (previous solvent not thoroughly flushed out)	**4f**	Flush system with compatible solvents until only the desired mobile phase is present
4g	Mobile phase not in equilibrium with column	**4g**	Continue to flush system until equilibrium is established
4h	Mobile phase/sample vaporizing	**4h**	At elevated temperatures check boiling point of mobile phase
4i	Contamination in mobile phase	**4i**	Change mobile phase
4j	Failing detector source	**4j**	Replace with new source
4k	Recorder problems	**4k**	Short out detector; if drift continues, check recorder

Symptom	Possible Cause	Corrective Action
5 Baseline stepping and peaks are flat-topped; Baseline does not zero	**5a** Recorder gain and damping improperly adjusted	**5a** Properly adjust gain and damping
	5b Improper grounding	**5b** Check system ground
	5c Saturated electronics	**5c** Reduce sample size
6 Baseline spiking	**6a** Air bubbles passing through detector	**6a1** Degas mobile phase
		6a2 Locate and repair all leaks
		6a3 Flush air out of pump and check valves
		6a4 Check boiling point of mobile phase
	6b Improper system ground	**6b** Check for proper grounding

Symptom	Cause		Remedy	
	6c	Electronic interference from other lab equipment turning on and off	**6c**	Check for other equipment turning on and off on same circuit and remove (e.g., constant temperature bath)
	6d	Loose electronic connections	**6d1**	Check all connections and plugs
			6d2	Check for vibrations
			6d3	Check for loose fitting source lamp
	6f	R_f feedback	**6f**	Properly ground equipment
7 Negative peaks	**7a**	Polarity reversed on detector or recorder	**7a1**	Switch polarity to other position
			7a2	Reverse detector leads
	7b	Negative peaks in UV trace	**7b**	Check for non-UV absorbers in system or sample impurities
	7c	Negative peak at V_O	**7c1**	Result of pressure surge due to sample introduction; Don't quantitate peaks at V_O
			7c2	Air sampled, improve sample introduction technique
8 Poor peak shape	**8a**	Flat bottomed peaks		
Flat bottom peaks	**8a1**	Bubbles in detector	**8a1**	Degas mobile phase or add suitable back pressure to detector cell
	8a2	Dirt buildup on detector cell windows	**8a2**	Clean detector cell
	8a3	Optics out of adjustment	**8a3**	Check alignment or call service representative
	8a4	Light output in reference beam lower than on sample side	**8a4**	Clean cell and check with operation manual or service representative

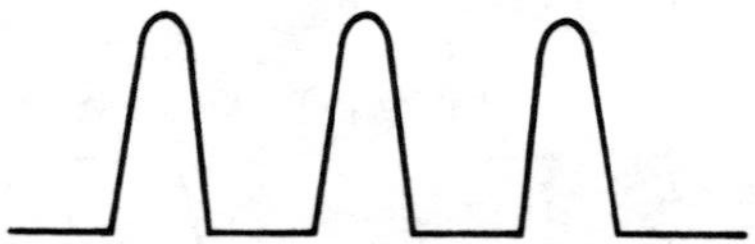

Symptom	Possible Cause	Corrective Action
 Very rounded peaks	**8b** Very rounded peaks	
	8b1 Operating beyond linear dynamic range of detector	**8b1** Reduce sample size
	8b2 Recorder gain is too low	**8b2** Adjust recorder gain
	8c Column-sample interaction (e.g., absorption)	**8c1** Check sample chemistry and change column **8c2** Increase temperature **8c3** Change ionic strength or pH
	8d Column dried out at ends	**8d** Replace column
	8e Column overload	**8e** Reduce sample size
	8f Contamination on detector cell windows	**8f** Clean detector cell
9 Loss of resolution	**9a** Column overload	**9a** Reduce sample size
	9b Loss of column efficiency	**9b** Adjust mobile phase or replace/regenerate column
	9c Loss of column liquid phase	**9c** Replace the column
	9d Dirty column	**9d** Wash column with suitable solvents or replace
	9e Distorted column bed (cracked, compressed)	**9e** Repack or replace column
	9f Used wrong column or mobile phase	**9f** Change system

Problem		Cause		Remedy	
10	Increased retention volume	**10a**	System flow rate decreased	**10a**	Check and increase flow rate. If flow rate decreasing, check and repair any leaks
		10b	Column temperature too low	**10b**	Insulate or jacket column
		10c	Column activity changing	**10c**	Solvent stripping H_2O or stationary phase off of column; add H_2O or liquid phase to mobile phase
11	Decreased retention time	**11a**	System flow rate increased	**11a**	Check pump for proper setting
		11b	Column activity changing	**11b**	Change column
		11c	Wrong mobile phase	**11c1**	Change mobile phase
				11c2	Check for changes in mobile phase ratio
12	Recorder will not zero	**12a**	Bubbles in UV reference cell or wrong solvent in RI reference cell	**12a1**	Flush cell out and replace with air or proper solvent
				12a2	Degas mobile phase
				12a3	Add suitable back pressure to detector outlet
		12b	Mobile phase has not equilibrated with column	**12b**	Flush system longer
		12c	Mobile phase has not equilibrated with detector or is not compatible with detector	**12c1**	The mobile phase's refractive index or UV is not compatible with the detector; adjust detector or change mobile phase
				12c2	Contaminant in mobile phase; prepare fresh and change, allowing time to equilibrate.

Symptom	Possible Cause	Corrective Action
	12d Column bleed	**12d1** Add H_2O or stationary phase to mobile phase
		12d2 Check solubility of stationary phase in mobile phase; if working above ambient, reduce temperature to check effect
	12e Contamination bleed from column	**12e** Thoroughly wash column with proper solvents or replace column
	12f Previous mobile phase still in system	**12f** Properly flush entire system with compatible solvents
	12g Detector not connected to recorder	**12g** Check detector lines to recorder
	12h Detector source lamp failing/faulty	**12h** Replace source
	12i Dirty detector cell windows	**12i** Clean cell windows
	12j Particulates in detector cell	**12j** Flush and clean detector cell
	12k Electronic problem with detector	**12k** Check service manual
	12l Recorder or detector not plugged in or turned on	**12l** Check power cord and on/off switch
	12m Recorder improperly zeroed	**12m** Rezero recorder
	12n Recorder calibration knob out of position	**12n** Readjust calibration knob

Problem		Cause		Remedy	
13	Low sensitivity	**13a**	Inadequate flow rate	**13a**	Adjust flow rate
		13b	Sample not compatible with detector	**13b**	Check sample chemistry and adjust detector or change it
		13c	Insufficient sample	**13c**	Increase sample size
		13d	Sample not eluting from column	**13d**	Check sample chemistry; change mobile phase and/or column
		13e	Dirty detector cell windows	**13e**	Clean cell
		13f	Gas bubble(s) in detector cell	**13f1**	Degas mobile phase
				13f2	Apply suitable back pressure to detector output
		13g	Detector attenuation too high	**13g**	Adjust attenuation
		13h	Detector and/or recorder out of calibration	**13h**	Check detector and recorder calibration; recalibrate if necessary
		13i	Failing/faulty detector source	**13i**	Change detector source
		13j	Recorder in wrong millivolt range	**13j**	Check setting and adjust
14	No peaks; no response	**14a**	Detector/recorder not on	**14a**	Check and turn on
		14b	Detector/recorder not plugged in	**14b**	Check and plug in
		14c	No sample injected	**14c**	Check injection/injector for complete sample introduction; clean or change syringe or valve
		14d	Failure in the electronics	**14d1**	Check and replace fuse
				14d2	Check service manual

BIBLIOGRAPHY

1 E. L. Johnson and R. Stevenson, *Basic Liquid Chromatography*, Varian Associates, Palo Alto, CA 1978.

2 L. R. Snyder and J. J. Kirkland, *Introduction to Modern Liquid Chromatography*, 2nd ed., John Wiley & Sons, Inc., New York, 1979.

3 J. Q. Walker, M. T. Jackson, Jr., and J. B. Maynard, *Chromatographic Systems—Maintenance and Troubleshooting*, 2nd ed., Academic Press, New York, 1977.

Appendix

2 Calibration of a Laboratory Strip Chart Recorder*

Perform this procedure at periodically scheduled intervals, for example, after every six months of use or at least once a year, and when new or taken out of storage for use.

1 Chart speed

a. Set the recorder chart speed selector to the speed most frequently used.
b. Turn on the chart drive.
c. Observe the chart paper, and when it is moving steadily, simultaneously start a timer and mark the chart paper with a small movement of the pen.
d. Allow the chart paper to run for at least 4–5 inches, but not less than 5 min of running time.
e. Mark the end of the test on the chart with another small movement of the pen.

*This is an example of a strip chart recorder calibration procedure and may not directly apply to your specific piece of equipment without some modifications. Consult the recorder's operation manual or technical service representative for specific procedures and/or specifications.

f. Stop the recorder and record the time interval on the chart paper.
g. Determine the actual distance traveled, and calculate the rate of speed.
h. The calculated rate should be within 2% of the nominal rate where the chart speed is an essential parameter of the data measurement.

2 **Gain and damping**

a. Set the recorder span or range switch to the most frequently used voltage.
b. Adjust the recorder zero to about 20% full scale deflection.
c. Apply a potential sufficient to move the pen to about 80% of full scale with a calibrated potential source such as a Dorhman A-100 microcoulometer/recorder analyzer or an Alltech Associates recorder check (part number 7039).
d. Turn the potential source on and off several times to produce a series of deflections between 20 and 80% of full scale.
e. There should not be excessive overshoot or undershoot. If needed, make the necessary gain and/or damping adjustments.

3 **Pen response**

a. Turn the calibrated potential source off (used in 2 above) and rezero the recorder to 0% of full scale using the same span or range setting as in 2 above.
b. Turn the potential source on and adjust it to give an approximate full scale reading.
c. Alternately switch the input from zero to full scale and back a few times and observe the pen response.
d. There should be no more than about 1 sec without excessive overshoot or undershoot.
e. Make any necessary gain and/or damping adjustments. If any adjustments are made, repeat step 2 to be sure the new settings are satisfactory. If proper adjustments with steps 2 and 3 cannot be made, call the technical service representative for the recorder.

4 **Dead band**

a. Apply a potential to the recorder input equal to about 50% full scale deflection with the recorder chart drive on.
b. Displace the pen quickly to the left one or two chart divisions and gently release it.
c. Allow the chart paper to run a short distance (about ½ inch) and then quickly displace the pen to the right one or two chart divisions and again gently release it.

d. Allow the chart paper to run a short distance again, and stop the chart paper.
e. Raise the pen.
f. Determine the distance between the two pen traces. This distance should be less than 1.0% of full scale. Repeat the procedure as necessary, making the required adjustments. If the gain and/or damping needs to be adjusted, repeat steps 2 and 3 to make sure that these adjustments are still satisfactory. If the adjustments for steps 2, 3, and 4 cannot be met simultaneously, call the technical service representative for the recorder.

5 Accuracy

a. Short out the recorder with the span or range switch set to the most frequently used voltage.
b. Adjust the zero control until the pen is at zero.
c. Apply a voltage to the recorder equal to the full scale voltage selected with a potential source, such as in step 2.
d. Adjust, if necessary, the span or range control to bring the recorder pen exactly to full scale. If the recorder has a continuously variable voltage input, be sure that this is in the calibrate or off position before running this test.

6 Linearity

a. Apply a minimum of five almost equally distributed known potentials from 0 to 100% of the full scale range selected with a calibrated potential source, such as in step 2, and with the recorder properly zeroed and calibrated from steps 1–5.
b. The recorder pen should indicate the known applied voltage to within ±1.0% of full scale at each level. If the linearity is not true, contact the technical service representative for the recorder.

7 Recording the calibration results

a. Record the following information using a suitable logbook:
   (1) The date of the test(s)
   (2) The name of the person(s) performing the test(s)
   (3) The values obtained for the
      (a) chart speed
      (b) linearity
      (c) average dead band
      (d) pen response and noise level
   (4) The recorder trace from these tests should be properly affixed to the logbook page and identified.

Appendix

3 Tools for the HPLC Laboratory

To perform most maintenance and troubleshooting procedures, the chromatographer will need a supply of quality tools. When planning or designing an HPLC laboratory, most users consider adequate hood space and duct work design, a sufficient number of separate electrical circuits, adequate bench space for instruments and sample preparation, and work space in order to work in front and behind the instruments. However, most do not include the necessary laboratory tools in the initial laboratory plan. This is important not only for the obvious repair purposes, but also to adequately define the laboratory startup costs.

Many tools the chromatographer will find useful are listed below.

Safety glasses with side shields
Dust mask
Toolbox or case
Parts cabinet with plenty of drawers (especially for storing HPLC fittings)
End wrenches, including metric sizes*

*(Wrenches with long lever arms (noted with asterisks in this list) are not recommended for most HPLC maintenance and repair work. Too much torque will easily gall HPLC fittings.

Allen-hex wrenches
Spline wrenches
Crescent wrenches*
Screwdrivers with standard slotted blades
Phillips head screwdrivers
Slip-joint pliers
Vise grip pliers
Long nose pliers
Curved needle nose pliers
Internal-external retaining ring pliers
Wire strippers
Wire cutter
File set and stainless steeel tube cutting kit
Flashlight
Tweezers
Pocket knife
Engraving tool
Dremel moto-tool and kit
Vise
Third hand or other holding jig
Soldering iron with stand and assortment of tips
Rosin solder
Volt-ohm meter
Ammeter
Assortment of screws, washers, nuts, and electrical connectors
Special tools recommended by the HPLC manufacturer

Other tools may be needed, and they can be purchased subsequently.

It is advised to keep the laboratory tools locked and/or engraved or color coded. This ensures that they will be available when needed and retrievable when lost or borrowed.

Appendix

4 Before Calling the Service Representative

Most manufacturers of HPLC equipment have technical and/or service representatives who are available to help the user solve application and equipment problems. The extent of this coverage is often a key selling point of the manufacturer as well as an important consideration by the buyer when selecting equipment.

No matter how good the manufacturer's intentions are, their representatives may not be immediately available by phone or visit, if necessary, when you call or need immediate help. For this reason, being able to handle simple equipment checks in your own laboratory can save you time and money. In addition, when you do call the representative, you can describe what already has been checked. This will help the manufacturer diagnose the problem sooner and not waste your time checking simple things.

The following list briefly describes many items the user should and can check before making the call. It by no means includes every possible alternative. However, it should aid in illustrating the direction to follow as well as serve as a model to develop your own check list. These as well as many more points are described in more detail throughout this text. Yet these simple checks, which should become part of the user's CPMA, are so frequently overlooked.

Instrument(s) on
Instrument(s) plugged in
Fuse(s) blown
Other equipment on the same electrical circuit
No mobile phase
Air block in the mobile phase reservoir-pump line
Pump pressure relief valve engaged
Gas bubbles in pump head(s)
Pump gas supply depleted
Broken pump plunger
Worn or leaking pump seal
Need a pulse dampener or restrictor
No sample being introduced
Solvent incompatible septum
Leaking septum
Leaking sample valve (e.g., crossport scratches on rotor)
Injection valve improperly positioned
Temperature gradient(s) across the system
Contaminated or plugged column
Wrong column
Flow through column reversed
Wrong detector for sample
Wrong or no detector reference
Dirty detector cell
Faulty or failing detector source
Wrong detector attenuation
Broken or cracked detector cell window(s)
System leak(s)
All indicator lights and gauges functioning properly
Wrong recorder or detector channel
Wrong recorder chart speed
Wrong recorder millivolt setting for detector
Recorder not grounded or shielded
Review sample's chemistry
Read instrument or equipment manual, including its troubleshooting guide

Index